IPMC

INTERNATIONAL
PROPERTY MAINTENANCE
CODE®

A Member of the International Code Family®

2021

INTERNATIONAL CODE COUNCIL®

2021 International Property Maintenance Code®

Date of First Publication: September 30, 2020

First Printing: September 2020

ISBN: 978-1-60983-971-0 (soft-cover edition)

T025354

PRINTED IN THE USA

PREFACE

Introduction

The *International Property Maintenance Code*® (IPMC®) establishes minimum requirements for the maintenance of existing buildings through model code regulations that contain clear and specific property maintenance and property improvement provisions. This 2021 edition is fully compatible with all of the International Codes® (I-Codes®) published by the International Code Council® (ICC®), including the *International Building Code*® (IBC®), *International Energy Conservation Code*® (IECC®), *International Existing Building Code*® (IEBC®), *International Fire Code*® (IFC®), *International Fuel Gas Code*® (IFGC®), *International Green Construction Code*® (IgCC®), *International Mechanical Code*® (IMC®), *International Plumbing Code*® (IPC®), *International Private Sewage Disposal Code*® (IPSDC®), *International Residential Code*® (IRC®), *International Swimming Pool and Spa Code*® (ISPSC®), *International Wildland-Urban Interface Code*® (IWUIC®), *International Zoning Code*® (IZC®) and *International Code Council Performance Code*® (ICCPC®).

The I-Codes, including the IPMC, are used in a variety of ways in both the public and private sectors. Most industry professionals are familiar with the I-Codes as the basis of laws and regulations in communities across the US and in other countries. However, the impact of the codes extends well beyond the regulatory arena, as they are used in a variety of nonregulatory settings, including:

- Voluntary compliance programs such as those promoting sustainability, energy efficiency and disaster resistance.
- The insurance industry, to estimate and manage risk, and as a tool in underwriting and rate decisions.
- Certification and credentialing of individuals involved in the fields of building design, construction and safety.
- Certification of building and construction-related products.
- US federal agencies, to guide construction in an array of government-owned properties.
- Facilities management.
- "Best practices" benchmarks for designers and builders, including those who are engaged in projects in jurisdictions that do not have a formal regulatory system or a governmental enforcement mechanism.
- College, university and professional school textbooks and curricula.
- Reference works related to building design and construction.

In addition to the codes themselves, the code development process brings together building professionals on a regular basis. It provides an international forum for discussion and deliberation about building design, construction methods, safety, performance requirements, technological advances and innovative products.

Development

This 2021 edition presents the code as originally issued, with changes reflected in the 2003 through 2018 editions and with further changes developed through the ICC Code Development Process through 2019. A new edition of the code is promulgated every 3 years.

This code is intended to establish provisions that adequately protect public health, safety and welfare and that neither unnecessarily increase construction costs nor give preferential treatment to particular types or classes of materials, products or methods of construction.

Maintenance

The *International Property Maintenance Code* is kept up to date through the review of proposed changes submitted by code enforcement officials, industry representatives, design professionals and other interested parties. Proposed changes are carefully considered through an open code development process in which all interested and affected parties may participate.

The ICC Code Development Process reflects principles of openness, transparency, balance, due process and consensus, the principles embodied in OMB Circular A-119, which governs the federal government's use of private-sector standards. The ICC process is open to anyone; there is no cost to participate, and people can participate without travel cost through the ICC's cloud-based app, cdpAccess®. A broad cross section of interests are represented in the ICC Code Development Process. The codes, which are updated regularly, include safeguards that allow for emergency action when required for health and safety reasons.

In order to ensure that organizations with a direct and material interest in the codes have a voice in the process, the ICC has developed partnerships with key industry segments that support the ICC's important public safety mission. Some code development committee members were nominated by the following industry partners and approved by the ICC Board:

- American Institute of Architects (AIA)
- National Association of Home Builders (NAHB)

The code development committees evaluate and make recommendations regarding proposed changes to the codes. Their recommendations are then subject to public comment and council-wide votes. The ICC's governmental members–public safety officials who have no financial or business interest in the outcome–cast the final votes on proposed changes.

The contents of this work are subject to change through the code development cycles and by any governmental entity that enacts the code into law. For more information regarding the code development process, contact the Codes and Standards Development Department of the ICC.

While the I-Code development procedure is thorough and comprehensive, the ICC, its members and those participating in the development of the codes disclaim any liability resulting from the publication or use of the I-Codes, or from compliance or noncompliance with their provisions. The ICC does not have the power or authority to police or enforce compliance with the contents of this code.

Code Development Committee Responsibilities (Letter Designations in Front of Section Numbers)

In each code development cycle, proposed changes to this code are considered at the Code Development Hearings by the International Property Maintenance/Zoning Code Development Committee, whose action constitutes a recommendation to the voting membership for final action on the proposed changes. Proposed changes to a code section having a number beginning with a letter in brackets are considered by a different code development committee. For example, proposed changes to code sections that have the letter [F] in front of them (e.g., [F] 704.1) are considered by the International Fire Code Development Committee at the Committee Action Hearing.

The content of sections in this code that begin with a letter designation is maintained by another code development committee in accordance with the following:

[A] = Administrative Code Development Committee;

[F] = International Fire Code Development Committee;

[P] = International Plumbing Code Development Committee;

[BE] = IBC–Egress Code Development Committee; and

[BG] = IBC–General Code Development Committee.

For the development of the 2024 edition of the I-Codes, there will be two groups of code development committees and they will meet in separate years, as shown in the following Code Development Hearings Table.

Code change proposals submitted for code sections that have a letter designation in front of them will be heard by the respective committee responsible for such code sections. Because different committees hold Committee Action Hearings in different years, proposals for the IPMC will be heard by committees in both the 2021 (Group A) and the 2022 (Group B) code development cycles.

For instance, every section of Chapter 1 of this code is designated as the responsibility of the Administrative Code Development Committee, which is part of the Group B portion of the hearings. This committee will hold its Committee Action Hearings in 2022 to consider code change proposals for Chapter 1 of all I-Codes except the IECC, IRC and IgCC. Therefore, any proposals received for Chapter 1 of this code will be assigned to the Administrative Code Development Committee for consideration in 2022.

It is very important that anyone submitting code change proposals understands which code development committee is responsible for the section of the code that is the subject of the code change proposal. For further information on the Code Development Committee responsibilities, please visit the ICC website at www.iccsafe.org/current-code-development-cycle.

CODE DEVELOPMENT HEARINGS

Group A Codes (Heard in 2021, Code Change Proposals Deadline: January 11, 2021)	Group B Codes (Heard in 2022, Code Change Proposals Deadline: January 10, 2022)
International Building Code– Egress (Chapters 10, 11, Appendix E) – Fire Safety (Chapters 7, 8, 9, 14, 26) – General (Chapters 2–6, 12, 27–33, Appendices A, B, C, D, K, N)	Administrative Provisions (Chapter 1 of all codes except IECC, IRC and IgCC; IBC Appendix O; the appendices titled "Board of Appeals" for all codes except IECC, IRC, IgCC, ICCPC and IZC; administrative updates to currently referenced standards; and designated definitions)
International Fire Code	**International Building Code** – Structural (Chapters 15–25, Appendices F, G, H, I, J, L, M)
International Fuel Gas Code	**International Existing Building Code**
International Mechanical Code	**International Energy Conservation Code—Commercial**
International Plumbing Code	**International Energy Conservation Code—Residential** – IECC—Residential – IRC—Energy (Chapter 11)
International Property Maintenance Code	**International Green Construction Code** (Chapter 1)
International Private Sewage Disposal Code	**International Residential Code** – IRC—Building (Chapters 1–10; Appendices AE, AF, AH, AJ, AK, AL, AM, AO, AQ, AR, AS, AT, AU, AV, AW)
International Residential Code – IRC—Mechanical (Chapters 12–23) – IRC—Plumbing (Chapters 25–33, Appendices AG, AI, AN, AP)	
International Swimming Pool and Spa Code	
International Wildland-Urban Interface Code	
International Zoning Code	

Note: Proposed changes to the ICCPC will be heard by the code development committee noted in brackets [ ] in the text of the ICCPC.

Marginal Markings

Solid vertical lines in the margins within the body of the code indicate a technical change from the requirements of the 2018 edition. Deletion indicators in the form of an arrow (➡) are provided in the margin where an entire section, paragraph, exception or table has been deleted or an item in a list of items or a row of a table has been deleted.

A single asterisk [*] placed in the margin indicates that text or a table has been relocated within the code. A double asterisk [**] placed in the margin indicates that the text or table immediately following it has been relocated there from elsewhere in the code. The following table indicates such relocations in the 2021 edition of the IPMC.

RELOCATIONS

2021 LOCATION	2018 LOCATION
104.1	103.5
105	104
105.7-105.7.1	103.4-103.4.1
106	105
107.2	111.1
108.1	111.2
109	106
110	112
111	108
111.4-111.6	107.1-107.4, 107.6
112	109
113	110
B101.2.2	111.8
B101.3	111.2
B101.3.2	111.2.1
B101.3.4	111.2.2
B101.3.5	111.2.4
B101.3.6	111.2.3
B101.3.7	111.2.5
B101.4	111.4.1
B101.5	111.3
B101.5.3	111.5
B101.7	111.6
B101.7.1	111.6.1
B101.8	111.7

Coordination of the International Codes

The coordination of technical provisions is one of the strengths of the ICC family of model codes. The codes can be used as a complete set of complementary documents, which will provide users with full integration and coordination of technical provisions. Individual codes can also be used in subsets or as stand-alone documents. To make sure that each individual code is as complete as possible, some technical provisions that are relevant to more than one subject area are duplicated in some of the model codes. This allows users maximum flexibility in their application of the I-Codes.

Italicized Terms

Terms italicized in code text, other than document titles, are defined in Chapter 2. The terms selected to be italicized have definitions that the user should read carefully to better understand the code. Where italicized, the Chapter 2 definition applies. If not italicized, common-use definitions apply.

Adoption

The ICC maintains a copyright in all of its codes and standards. Maintaining copyright allows the ICC to fund its mission through sales of books, in both print and electronic formats. The ICC welcomes adoption of its codes by jurisdictions that recognize and acknowledge the ICC's copyright in the code, and further acknowledge the substantial shared value of the public/private partnership for code development between jurisdictions and the ICC.

The ICC also recognizes the need for jurisdictions to make laws available to the public. All I-Codes and I-Standards, along with the laws of many jurisdictions, are available for free in a nondownloadable form on the ICC's website. Jurisdictions should contact the ICC at adoptions@iccsafe.org to learn how to adopt and distribute laws based on the IPMC in a manner that provides necessary access, while maintaining the ICC's copyright.

To facilitate adoption, several sections of this code contain blanks for fill-in information that needs to be supplied by the adopting jurisdiction as part of the adoption legislation. For this code, please see:

Section 101.1. Insert: **[NAME OF JURISDICTION]**

Section 103.1. Insert: **[NAME OF DEPARTMENT]**

Section 302.4. Insert: **[HEIGHT IN INCHES]**

Section 304.14. Insert: **[DATES IN TWO LOCATIONS]**

Section 602.3. Insert: **[DATES IN TWO LOCATIONS]**

Section 602.4. Insert: **[DATES IN TWO LOCATIONS]**

Effective Use of the International Property Maintenance Code

The IPMC is a model code that regulates the minimum maintenance requirements for existing buildings.

The IPMC is a maintenance document intended to establish minimum maintenance standards for basic equipment, light, ventilation, heating, sanitation and fire safety. Responsibility is fixed among owners, operators and occupants for code compliance. The IPMC provides for the regulation and safe use of existing structures in the interest of the social and economic welfare of the community.

ARRANGEMENT AND FORMAT OF THE 2021 IPMC

Before applying the requirements of the IPMC, it is beneficial to understand its arrangement and format. The IPMC, like other codes published by ICC, is arranged and organized to follow sequential steps that generally occur during a plan review or inspection. Below is a chapter-by-chapter synopsis of the scope and intent of the provisions of the IPMC. The following table shows how the IPMC is divided. The ensuing chapter-by-chapter synopsis details the scope and intent of the provisions of the IPMC.

Chapter Topics

Chapter	Subjects
1	Scope and Administration
2	Definitions
3	General Requirements
4	Light, Ventilation and Occupancy Limitations
5	Plumbing Facilities and Fixture Requirements
6	Mechanical and Electrical Requirements
7	Fire Safety Requirements
8	Referenced Standards
Appendix A	Boarding Standard
Appendix B	Board of Appeals

Chapter 1 Scope and Administration

Chapter 1 contains provisions for the application, enforcement and administration of subsequent requirements of the code. In addition to establishing the scope of the code, Chapter 1 identifies which buildings and structures come under its purview. Chapter 1 is largely concerned with maintaining "due process of law" in enforcing the property maintenance criteria contained in the body of the code. Only through careful observation of the administrative provisions can the building official reasonably expect to demonstrate that "equal protection under the law" has been provided.

Chapter 2 Definitions

All terms that are defined in the code are listed alphabetically in Chapter 2. While a defined term may be used in one chapter or another, the meaning provided in Chapter 2 is applicable throughout the code.

Where understanding of a term's definition is especially key to or necessary for understanding of a particular code provision, the term is shown in italics. This is true only for those terms that have a meaning that is unique to the code. In other words, the generally understood meaning of a term or phrase might not be sufficient or consistent with the meaning prescribed by the code; therefore, it is essential that the code-defined meaning be known.

Guidance is provided regarding tense, gender and plurality of defined terms as well as guidance regarding terms not defined in this code.

Chapter 3 General Requirements

Chapter 3, "General Requirements," is broad in scope. It includes a variety of requirements for the exterior property areas as well as the interior and exterior elements of the structure. This chapter provides requirements that are intended to maintain a minimum level of safety and sanitation for both the general public and the occupants of a structure, and to maintain a building's structural and weather-resistance performance. Chapter 3 provides specific criteria for regulating the installation and maintenance of specific building components; maintenance requirements for vacant structures and land; requirements regulating the safety, sanitation and appearance of the interior and exterior of structures and all exterior property areas; accessory structures; vehicle storage regulations and establishes who is responsible for complying with the chapter's provisions. This chapter also contains the requirements for swimming pools, spas and hot tubs and the requirements for protective barriers and gates in these barriers. Chapter 3 establishes the responsible parties for exterminating insects and rodents, and maintaining sanitary conditions in all types of occupancies.

Chapter 4 Light, Ventilation and Occupancy Limitations

The purposes of Chapter 4 are to set forth these requirements in the code and to establish the minimum environment for occupiable and habitable buildings, by establishing the minimum criteria for light and ventilation and identifying occupancy limitations including minimum room width and area, minimum ceiling height and restrictions to prevent overcrowding. This chapter also provides for alternative arrangements of windows and other devices to comply with the requirements for light and ventilation and prohibits certain room arrangements and occupancy uses.

Chapter 5 Plumbing Facilities and Fixture Requirements

Chapter 5 establishes the minimum criteria for the installation, maintenance and location of plumbing systems and facilities, including the water supply system, water heating appliances, sewage disposal system and related plumbing fixtures.

Sanitary and clean conditions in occupied buildings are dependent upon certain basic plumbing principles, including providing potable water to a building, providing the basic fixtures to effectively utilize that water and properly removing waste from the building. Chapter 5 establishes the minimum criteria to verify that these principles are maintained throughout the life of a building.

Chapter 6 Mechanical and Electrical Requirements

The purpose of Chapter 6 is to establish minimum performance requirements for heating, electrical and mechanical facilities and to establish minimum standards for the safety of these facilities.

This chapter establishes minimum criteria for the installation and maintenance of the following: heating and air-conditioning equipment, appliances and their supporting systems; water heating equipment, appliances and systems; cooking equipment and appliances; ventilation and exhaust equipment; gas and liquid fuel distribution piping and components; fireplaces and solid fuel-burning appliances; chimneys and vents; electrical services; lighting fixtures; electrical receptacle outlets; electrical distribution system equipment, devices and wiring; and elevators, escalators and dumbwaiters.

Chapter 7 Fire Safety Requirements

The purpose of Chapter 7 is to address those fire hazards that arise as the result of a building's occupancy. It also provides minimum requirements for fire safety issues that are most likely to arise in older buildings.

This chapter contains requirements for means of egress in existing buildings, including path of travel, required egress width, means of egress doors and emergency escape openings.

Chapter 8 Referenced Standards

The code contains numerous references to standards that are used to regulate materials and methods of construction. Chapter 8 contains a comprehensive list of all standards that are referenced in the code. The standards are part of the code to the extent of the reference to the standard. Compliance with the referenced standard is necessary for compliance with this code. By providing specifically adopted standards, the construction and installation requirements necessary for compliance with the code can be readily determined. The basis for code compliance is, therefore, established and available on an equal basis to the code official, contractor, designer and owner.

Chapter 8 is organized in a manner that makes it easy to locate specific standards. It lists all of the referenced standards, alphabetically, by acronym of the promulgating agency of the standard. Each agency's standards are then listed in either alphabetical or numeric order based upon the standard identification. The list also contains the title of the standard; the edition (date) of the standard referenced; any addenda included as part of the ICC adoption; and the section or sections of this code that reference the standard.

Appendix A Boarding Standard

The provisions in Appendix A are not mandatory unless specifically referenced in the adopting ordinance. This appendix provides minimum specifications for boarding a structure. This can be utilized by a jurisdiction as a set of minimum requirements in order to result in consistent boarding quality.

Appendix B Board of Appeals

The provisions contained in Appendix B are not mandatory unless specifically referenced in the adopting ordinance. This appendix provides criteria for Board of Appeals members and procedures by which the Board of Appeals should conduct its business.

TABLE OF CONTENTS

CHAPTER 1
SCOPE AND ADMINISTRATION

User note:

About this chapter: *Chapter 1 establishes the limits of applicability of the code and describes how the code is to be applied and enforced. Chapter 1 is in two parts: Part 1—Scope and Application (Sections 101 and 102) and Part 2—Administration and Enforcement (Sections 103 – 110). Section 101 identifies which buildings and structures come under its purview and references other I-Codes as applicable.*

This code is intended to be adopted as a legally enforceable document and it cannot be effective without adequate provisions for its administration and enforcement. The provisions of Chapter 1 establish the authority and duties of the code official appointed by the authority having jurisdiction and also establish the rights and privileges of the property owner and building occupants.

This Chapter was extensively reorganized for the 2021 edition. For clarity, the relocation marginal markings have not been included. For complete information, see the relocations table in the Preface information of this code.

PART 1 — SCOPE AND APPLICATION

SECTION 101
SCOPE AND GENERAL REQUIREMENTS

[A] 101.1 Title. These regulations shall be known as the *International Property Maintenance Code* of **[NAME OF JURISDICTION]**, hereinafter referred to as "this code."

[A] 101.2 Scope. The provisions of this code shall apply to all existing residential and nonresidential structures and all existing *premises* and constitute minimum requirements and standards for *premises*, structures, equipment and facilities for light, *ventilation,* space, heating, sanitation, protection from the elements, a reasonable level of safety from fire and other hazards, and for a reasonable level of sanitary maintenance; the responsibility of *owners,* an *owner*'s authorized agent, *operators* and *occupants;* the *occupancy* of existing structures and *premises,* and for administration, enforcement and penalties.

[A] 101.3 Purpose. The purpose of this code is to establish minimum requirements to provide a reasonable level of health, safety, property protection and general welfare insofar as they are affected by the continued *occupancy* and maintenance of structures and *premises*. Existing structures and *premises* that do not comply with these provisions shall be altered or repaired to provide a reasonable minimum level of health, safety and general welfare as required herein.

[A] 101.4 Severability. If a section, subsection, sentence, clause or phrase of this code is, for any reason, held to be unconstitutional, such decision shall not affect the validity of the remaining portions of this code.

SECTION 102
APPLICABILITY

[A] 102.1 General. Where there is a conflict between a general requirement and a specific requirement, the specific requirement shall govern. Where differences occur between provisions of this code and the referenced standards, the provisions of this code shall apply. Where, in a specific case, different sections of this code specify different requirements, the most restrictive shall govern.

102.2 Maintenance. Equipment, systems, devices and safeguards required by this code or a previous regulation or code under which the *structure* or *premises* was constructed, altered or repaired shall be maintained in good working order. An *owner, owner*'s authorized agent, *operator* or *occupant* shall not cause any service, facility, equipment or utility that is required under this section to be removed from, shut off from or discontinued for any occupied dwelling, except for such temporary interruption as necessary while repairs or alterations are in progress. The requirements of this code are not intended to provide the basis for removal or abrogation of fire protection and safety systems and devices in existing structures. Except as otherwise specified herein, the *owner* or the *owner's* authorized agent shall be responsible for the maintenance of buildings, structures and *premises.*

[A] 102.3 Application of other codes. Repairs, additions or alterations to a *structure*, or changes of *occupancy*, shall be done in accordance with the procedures and provisions of the *International Building Code, International Existing Building Code, International Energy Conservation Code, International Fire Code, International Fuel Gas Code, International Mechanical Code, International Residential Code, International Plumbing Code* and NFPA 70. Nothing in this code shall be construed to cancel, modify or set aside any provision of the *International Zoning Code*.

[A] 102.4 Existing remedies. The provisions in this code shall not be construed to abolish or impair existing remedies of the jurisdiction or its officers or agencies relating to the removal or demolition of any *structure* that is dangerous, unsafe and insanitary.

[A] 102.5 Workmanship. Repairs, maintenance work, alterations or installations that are caused directly or indirectly by the enforcement of this code shall be executed and installed in a *workmanlike* manner and installed in accordance with the manufacturer's instructions.

102.6 Structural analysis. Where structural analysis is used to determine if an unsafe structural condition exists, the analysis shall be permitted to use nominal strengths, nominal loads, load effects, required strengths and limit states in accordance with the requirements under which the *structure* was constructed or in accordance with any subsequent requirement.

[A] 102.7 Historic buildings. The provisions of this code shall not be mandatory for existing buildings or structures designated as historic buildings where such buildings or structures are judged by the *code official* to be safe and in the public interest of health, safety and welfare.

[A] 102.8 Referenced codes and standards. The codes and standards referenced in this code shall be those that are listed in Chapter 8 and considered part of the requirements of this code to the prescribed extent of each such reference and as further regulated in Sections 102.8.1 and 102.8.2.

Exception: Where enforcement of a code provision would violate the conditions of the listing of the equipment or appliance, the conditions of the listing shall apply.

[A] 102.8.1 Conflicts. Where conflicts occur between provisions of this code and the referenced standards, the provisions of this code shall apply.

[A] 102.8.2 Provisions in referenced codes and standards. Where the extent of the reference to a referenced code or standard includes subject matter that is within the scope of this code, the provisions of this code, as applicable, shall take precedence over the provisions in the referenced code or standard.

[A] 102.9 Requirements not covered by code. Requirements necessary for the strength, stability or proper operation of an existing fixture, *structure* or equipment, or for the public safety, health and general welfare, not specifically covered by this code, shall be determined by the *code official*.

[A] 102.10 Application of references. References to chapter or section numbers, or to provisions not specifically identified by number, shall be construed to refer to such chapter, section or provision of this code.

[A] 102.11 Other laws. The provisions of this code shall not be deemed to nullify any provisions of local, state or federal law.

PART 2 — ADMINISTRATION AND ENFORCEMENT

SECTION 103
CODE COMPLIANCE AGENCY

[A] 103.1 Creation of agency. The [INSERT NAME OF DEPARTMENT] is hereby created and the official in charge thereof shall be known as the code official. The function of the agency shall be the implementation, administration and enforcement of the provisions of this code.

[A] 103.2 Appointment. The *code official* shall be appointed by the chief appointing authority of the jurisdiction.

[A] 103.3 Deputies. In accordance with the prescribed procedures of this jurisdiction and with the concurrence of the appointing authority, the *code official* shall have the authority to appoint a deputy code official, other related technical officers, inspectors and other employees. Such employees shall have powers as delegated by the *code official*.

SECTION 104
FEES

[A] 104.1 Fees. The fees for activities and services performed by the department in carrying out its responsibilities under this code shall be as established by the applicable governing authority.

104.2 Refunds. The *code official* is authorized to establish a refund policy.

SECTION 105
DUTIES AND POWERS OF THE CODE OFFICIAL

[A] 105.1 General. The *code official* is hereby authorized and directed to enforce the provisions of this code. The *code official* shall have the authority to render interpretations of this code and to adopt policies and procedures in order to clarify the application of its provisions. Such interpretations, policies and procedures shall be in compliance with the intent and purpose of this code. Such policies and procedures shall not have the effect of waiving requirements specifically provided for in this code.

[A] 105.2 Inspections. The *code official* shall make all of the required inspections, or shall accept reports of inspection by *approved* agencies or individuals. Reports of such inspections shall be in writing and be certified by a responsible officer of such *approved* agency or by the responsible individual. The *code official* is authorized to engage such expert opinion as deemed necessary to report on unusual technical issues that arise, subject to the approval of the appointing authority.

[A] 105.3 Right of entry. Where it is necessary to make an inspection to enforce the provisions of this code, or whenever the *code official* has reasonable cause to believe that there exists in a *structure* or upon a *premises* a condition in violation of this code, the *code official* is authorized to enter the *structure* or *premises* at reasonable times to inspect or perform the duties imposed by this code, provided that if such *structure* or *premises* is occupied the *code official* shall present credentials to the *occupant* and request entry. If such *structure* or *premises* is unoccupied, the *code official* shall first make a reasonable effort to locate the *owner*, *owner*'s authorized agent or other person having charge or control of the *structure* or *premises* and request entry. If entry is refused, the *code official* shall have recourse to the remedies provided by law to secure entry.

[A] 105.4 Identification. The *code official* shall carry proper identification when inspecting *structures* or *premises* in the performance of duties under this code.

[A] 105.5 Notices and orders. The *code official* shall issue all necessary notices or orders to ensure compliance with this code.

[A] 105.6 Department records. The *code official* shall keep official records of all business and activities of the department specified in the provisions of this code. Such records shall be retained in the official records for the period required for retention of public records.

[A] 105.7 Liability. The *code official*, member of the board of appeals or employee charged with the enforcement of this code, while acting for the jurisdiction, in good faith and without malice in the discharge of the duties required by this code or other pertinent law or ordinance, shall not thereby be rendered civilly or criminally liable personally, and is hereby relieved from all personal liability for any damage accruing to persons or property as a result of an act or by reason of an act or omission in the discharge of official duties.

[A] 105.7.1 Legal defense. Any suit or criminal complaint instituted against any officer or employee because of an act performed by that officer or employee in the lawful discharge of duties and under the provisions of this code shall be defended by the legal representative of the jurisdiction until the final termination of the proceedings. The *code official* or any subordinate shall not be liable for costs in an action, suit or proceeding that is instituted in pursuance of the provisions of this code.

SECTION 106
APPROVAL

[A] 106.1 Modifications. Whenever there are practical difficulties involved in carrying out the provisions of this code, the *code official* shall have the authority to grant modifications for individual cases upon application of the *owner* or *owner's* authorized agent, provided that the *code official* shall first find that special individual reason makes the strict letter of this code impractical, the modification is in compliance with the intent and purpose of this code and that such modification does not lessen health, life and fire safety requirements. The details of action granting modifications shall be recorded and entered in the department files.

[A] 106.2 Alternative materials, design and methods of construction and equipment. The provisions of this code are not intended to prevent the installation of any material or to prohibit any design or method of construction not specifically prescribed by this code, provided that any such alternative has been *approved*. An alternative material, design or method of construction shall be *approved* where the *code official* finds that the proposed design is satisfactory and complies with the intent of the provisions of this code, and that the material, method or work offered is, for the purpose intended, not less than the equivalent of that prescribed in this code in quality, strength, effectiveness, fire resistance, durability and safety. Where the alternative material, design or method of construction is not *approved*, the *code official* shall respond in writing, stating the reasons why the alternative was not *approved*.

[A] 106.3 Required testing. Whenever there is insufficient evidence of compliance with the provisions of this code or evidence that a material or method does not conform to the requirements of this code, or in order to substantiate claims for alternative materials or methods, the *code official* shall have the authority to require tests to be made as evidence of compliance without expense to the jurisdiction.

[A] 106.3.1 Test methods. Test methods shall be as specified in this code or by other recognized test standards. In the absence of recognized and accepted test methods, the *code official* shall be permitted to approve appropriate testing procedures performed by an *approved* agency.

[A] 106.3.2 Test reports. Reports of tests shall be retained by the *code official* for the period required for retention of public records.

[A] 106.4 Used material and equipment. Materials that are reused shall comply with the requirements of this code for new materials. Materials, equipment and devices shall not be reused unless such elements are in good repair or have been reconditioned and tested where necessary, placed in good and proper working condition and *approved* by the *code official*.

[A] 106.5 Approved materials and equipment. Materials, equipment and devices *approved* by the *code official* shall be constructed and installed in accordance with such approval.

[A] 106.6 Research reports. Supporting data, where necessary to assist in the approval of materials or assemblies not specifically provided for in this code, shall consist of valid research reports from *approved* sources.

SECTION 107
MEANS OF APPEAL

107.1 General. In order to hear and decide appeals of orders, decisions or determinations made by the *code official* relative to the application and interpretation of this code, there shall be and is hereby created a board of appeals. The board of appeals shall be appointed by the applicable governing authority and shall hold office at its pleasure. The board shall adopt rules of procedure for conducting its business and shall render all decisions and findings in writing to the appellant with a duplicate copy to the *code official*.

[A] 107.2 Limitations of authority. An application for appeal shall be based on a claim that the true intent of this code or the rules legally adopted thereunder have been incorrectly interpreted, the provisions of this code do not fully apply or an equivalent or better form of construction is proposed. The board shall not have authority to waive requirements of this code or interpret the administration of this code.

107.3 Qualifications. The board of appeals shall consist of members who are qualified by experience and training and are not employees of the jurisdiction.

107.4 Administration. The *code official* shall take immediate action in accordance with the decision of the board.

SECTION 108
BOARD OF APPEALS

[A] 108.1 Membership of board. The board of appeals shall consist of not less than three members who are qualified by experience and training to pass on matters pertaining to property maintenance and who are not employees of the jurisdiction. The *code official* shall be an ex-officio member but shall not vote on any matter before the board. The board shall be appointed by the chief appointing authority, and shall serve staggered and overlapping terms.

SECTION 109 VIOLATIONS

[A] 109.1 Unlawful acts. It shall be unlawful for a person, firm or corporation to be in conflict with or in violation of any of the provisions of this code.

[A] 109.2 Notice of violation. The *code official* shall serve a notice of violation or order in accordance with Section 111.4.

[A] 109.3 Prosecution of violation. Any person failing to comply with a notice of violation or order served in accordance with Section 111.4 shall be deemed guilty of a misdemeanor or civil infraction as determined by the local municipality, and the violation shall be deemed a *strict liability offense*. If the notice of violation is not complied with, the *code official* shall institute the appropriate proceeding at law or in equity to restrain, correct or abate such violation, or to require the removal or termination of the unlawful *occupancy* of the *structure* in violation of the provisions of this code or of the order or direction made pursuant thereto. Any action taken by the authority having jurisdiction on such *premises* shall be charged against the real estate upon which the *structure* is located and shall be a lien upon such real estate.

[A] 109.4 Violation penalties. Any person who shall violate a provision of this code, or fail to comply therewith, or with any of the requirements thereof, shall be prosecuted within the limits provided by state or local laws. Each day that a violation continues after due notice has been served shall be deemed a separate offense.

[A] 109.5 Abatement of violation. The imposition of the penalties herein prescribed shall not preclude the legal officer of the jurisdiction from instituting appropriate action to restrain, correct or abate a violation, or to prevent illegal *occupancy* of a building, *structure* or *premises*, or to stop an illegal act, conduct, business or utilization of the building, *structure* or *premises*.

SECTION 110 STOP WORK ORDER

[A] 110.1 Authority. Where the *code official* finds any work regulated by this code being performed in a manner contrary to the provisions of this code or in a dangerous or unsafe manner, the *code official* is authorized to issue a stop work order.

[A] 110.2 Issuance. The stop work order shall be in writing and shall be given to the *owner* of the property, to the *owner's* authorized agent, or to the person performing the work. Upon issuance of a stop work order, the cited work shall immediately cease. The stop work order shall state the reason for the order and the conditions under which the cited work is authorized to resume.

[A] 110.3 Emergencies. Where an emergency exists, the *code official* shall not be required to give a written notice prior to stopping the work.

[A] 110.4 Failure to comply. Any person who shall continue any work after having been served with a stop work order, except such work as that person is directed to perform to remove a violation or unsafe condition, shall be subject to fines established by the authority having jurisdiction.

SECTION 111 UNSAFE STRUCTURES AND EQUIPMENT

111.1 Unsafe conditions. When a *structure* or equipment is found by the *code official* to be unsafe, or when a *structure* is found unfit for human *occupancy*, or is found unlawful, such *structure* shall be *condemned* pursuant to the provisions of this code.

111.1.1 Unsafe structures. An unsafe *structure* is one that is found to be dangerous to the life, health, property or safety of the public or the *occupants* of the *structure* by not providing minimum safeguards to protect or warn *occupants* in the event of fire, or because such *structure* contains unsafe equipment or is so damaged, decayed, dilapidated, structurally unsafe or of such faulty construction or unstable foundation, that partial or complete collapse is possible.

111.1.2 Unsafe equipment. Unsafe equipment includes any boiler, heating equipment, elevator, moving stairway, electrical wiring or device, flammable liquid containers or other equipment on the *premises* or within the *structure* that is in such disrepair or condition that such equipment is a hazard to life, health, property or safety of the public or *occupants* of the *premises* or *structure*.

111.1.3 Structure unfit for human occupancy. A *structure* is unfit for human *occupancy* whenever the *code official* finds that such *structure* is unsafe, unlawful or, because of the degree to which the *structure* is in disrepair or lacks maintenance, is insanitary, vermin or rat infested, contains filth and contamination, or lacks *ventilation*, illumination, sanitary or heating facilities or other essential equipment required by this code, or because the location of the *structure* constitutes a hazard to the *occupants* of the *structure* or to the public.

111.1.4 Unlawful structure. An unlawful *structure* is one found in whole or in part to be occupied by more persons than permitted under this code, or was erected, altered or occupied contrary to law.

111.1.5 Dangerous structure or premises. For the purpose of this code, any *structure* or *premises* that has any or all of the conditions or defects described as follows shall be considered to be dangerous:

1. Any door, aisle, passageway, stairway, exit or other means of egress that does not conform to the *approved* building or fire code of the jurisdiction

as related to the requirements for existing buildings.

2. The walking surface of any aisle, passageway, stairway, exit or other means of egress is so warped, worn loose, torn or otherwise unsafe as to not provide safe and adequate means of egress.
3. Any portion of a building, *structure* or appurtenance that has been damaged by fire, earthquake, wind, flood, *deterioration, neglect,* abandonment, vandalism or by any other cause to such an extent that it is likely to partially or completely collapse, or to become *detached* or dislodged.
4. Any portion of a building, or any member, appurtenance or ornamentation on the exterior thereof that is not of sufficient strength or stability, or is not so *anchored*, attached or fastened in place so as to be capable of resisting natural or artificial loads of one and one-half the original designed value.
5. The building or *structure,* or part of the building or *structure*, because of dilapidation, *deterioration,* decay, faulty construction, the removal or movement of some portion of the ground necessary for the support, or for any other reason, is likely to partially or completely collapse, or some portion of the foundation or underpinning of the building or *structure* is likely to fail or give way.
6. The building or *structure,* or any portion thereof, is clearly unsafe for its use and *occupancy*.
7. The building or *structure* is *neglected*, damaged, dilapidated, unsecured or abandoned so as to become an attractive nuisance to children who might play in the building or *structure* to their danger, becomes a harbor for vagrants, criminals or immoral persons, or enables persons to resort to the building or *structure* for committing a nuisance or an unlawful act.
8. Any building or *structure* has been constructed, exists or is maintained in violation of any specific requirement or prohibition applicable to such building or *structure* provided by the *approved* building or fire code of the jurisdiction, or of any law or ordinance to such an extent as to present either a substantial risk of fire, building collapse or any other threat to life and safety.
9. A building or *structure,* used or intended to be used for dwelling purposes, because of inadequate maintenance, dilapidation, decay, damage, faulty construction or arrangement, inadequate light, *ventilation,* mechanical or plumbing system, or otherwise, is determined by the *code official* to be unsanitary, unfit for human habitation or in such a condition that is likely to cause sickness or disease.
10. Any building or *structure,* because of a lack of sufficient or proper fire-resistance-rated construction, fire protection systems, electrical system, fuel connections, mechanical system, plumbing system or other cause, is determined by the *code official* to be a threat to life or health.
11. Any portion of a building remains on a site after the demolition or destruction of the building or *structure* or whenever any building or *structure* is abandoned so as to constitute such building or portion thereof as an attractive nuisance or hazard to the public.

111.2 Closing of vacant structures. If the *structure* is vacant and unfit for human habitation and *occupancy,* and is not in danger of structural collapse, the *code official* is authorized to post a placard of condemnation on the *premises* and order the *structure* closed up so as not to be an attractive nuisance. Upon failure of the *owner* or *owner*'s authorized agent to close up the *premises* within the time specified in the order, the *code official* shall cause the *premises* to be closed and secured through any available public agency or by contract or arrangement by private persons and the cost thereof shall be charged against the real estate upon which the *structure* is located and shall be a lien upon such real estate and shall be collected by any other legal resource.

111.2.1 Authority to disconnect service utilities. The *code official* shall have the authority to authorize disconnection of utility service to the building, *structure* or system regulated by this code and the referenced codes and standards set forth in Section 102.8 in case of emergency where necessary to eliminate an immediate hazard to life or property or where such utility connection has been made without approval. The *code official* shall notify the serving utility and, whenever possible, the *owner* or *owner*'s authorized agent and *occupant* of the building, *structure* or service system of the decision to disconnect prior to taking such action. If not notified prior to disconnection the *owner*, *owner*'s authorized agent or *occupant* of the building *structure* or service system shall be notified in writing as soon as practical thereafter.

111.3 Record. The *code official* shall cause a report to be filed on an unsafe condition. The report shall state the *occupancy* of the *structure* and the nature of the unsafe condition.

111.4 Notice. Whenever the *code official* determines that there has been a violation of this code or has grounds to believe that a violation has occurred, notice shall be given in the manner prescribed in Sections 111.4.1 and 111.4.2 to the owner or the owner's authorized agent, for the violation as specified in this code. Notices for condemnation procedures shall comply with this section.

111.4.1 Form. Such notice shall be in accordance with all of the following:

1. Be in writing.
2. Include a description of the real estate sufficient for identification.
3. Include a statement of the violation or violations and why the notice is being issued.
4. Include a correction order allowing a reasonable time to make the repairs and improvements required to bring the *dwelling unit* or *structure* into compliance with the provisions of this code.

5. Inform the property *owner* or *owner*'s authorized agent of the right to appeal.
6. Include a statement of the right to file a lien in accordance with Section 109.3.

111.4.2 Method of service. Such notice shall be deemed to be properly served where a copy thereof is served in accordance with one of the following methods:

1. A copy is delivered personally.
2. A copy is sent by certified or registered mail addressed to the owner at the last known address with the return receipt requested.
3. A copy is delivered in any other manner as prescribed by local law.

If the certified or registered letter is returned showing that the letter was not delivered, a copy thereof shall be posted in a conspicuous place in or about the structure affected by such notice. Service of such notice in the foregoing manner upon the owner's agent or upon the person responsible for the structure shall constitute service of notice upon the owner.

111.5 Unauthorized tampering. Signs, tags or seals posted or affixed by the *code official* shall not be mutilated, destroyed or tampered with, or removed without authorization from the *code official.*

111.6 Transfer of ownership. It shall be unlawful for the *owner* of any *dwelling unit* or *structure* who has received a compliance order or upon whom a notice of violation has been served to sell, transfer, mortgage, lease or otherwise dispose of such *dwelling unit* or *structure* to another until the provisions of the compliance order or notice of violation have been complied with, or until such *owner* or the *owner*'s authorized agent shall first furnish the grantee, transferee, mortgagee or lessee a true copy of any compliance order or notice of violation issued by the *code official* and shall furnish to the *code official* a signed and notarized statement from the grantee, transferee, mortgagee or lessee, acknowledging the receipt of such compliance order or notice of violation and fully accepting the responsibility without condition for making the corrections or repairs required by such compliance order or notice of violation.

111.7 Placarding. Upon failure of the *owner, owner*'s authorized agent or person responsible to comply with the notice provisions within the time given, the *code official* shall post on the *premises* or on defective equipment a placard bearing the word "Condemned" and a statement of the penalties provided for occupying the *premises,* operating the equipment or removing the placard. Such notice shall be posted in a conspicuous place in or about the structure affected by such notice. If the notice pertains to equipment, it shall be placed on the condemned equipment.

111.7.1 Placard removal. The *code official* shall remove the condemnation placard whenever the defect or defects upon which the condemnation and placarding action were based have been eliminated. Any person who defaces or removes a condemnation placard without the approval of the *code official* shall be subject to the penalties provided by this code.

111.8 Prohibited occupancy. Any occupied *structure condemned* and placarded by the *code official* shall be vacated as ordered by the *code official.* Any person who shall occupy a placarded *premises* or shall operate placarded equipment, and any *owner or owner*'s authorized agent who shall let anyone occupy a placarded *premises* or operate placarded equipment shall be liable for the penalties provided by this code.

111.9 Restoration or abatement. The structure or equipment determined to be unsafe by the code official is permitted to be restored to a safe condition. The *owner, owner*'s authorized agent, *operator* or *occupant* of a structure, *premises* or equipment deemed unsafe by the *code official* shall abate or cause to be abated or corrected such unsafe conditions either by repair, rehabilitation, demolition or other *approved* corrective action. To the extent that repairs, alterations, or additions are made or a change of occupancy occurs during the restoration of the structure, such repairs, alterations, additions, or change of occupancy shall comply with the requirements of the *International Existing Building Code.*

SECTION 112
EMERGENCY MEASURES

112.1 Imminent danger. When, in the opinion of the *code official,* there is *imminent danger* of failure or collapse of a building or *structure* that endangers life, or when any *structure* or part of a *structure* has fallen and life is endangered by the occupation of the *structure,* or when there is actual or potential danger to the building *occupants* or those in the proximity of any *structure* because of explosives, explosive fumes or vapors or the presence of toxic fumes, gases or materials, or operation of defective or dangerous equipment, the *code official* is hereby authorized and empowered to order and require the *occupants* to vacate the *premises* forthwith. The *code official* shall cause to be posted at each entrance to such *structure* a notice reading as follows: "This Structure Is Unsafe and Its Occupancy Has Been Prohibited by the Code Official." It shall be unlawful for any person to enter such *structure* except for the purpose of securing the structure, making the required repairs, removing the hazardous condition or of demolishing the same.

112.2 Temporary safeguards. Notwithstanding other provisions of this code, whenever, in the opinion of the *code official,* there is *imminent danger* due to an unsafe condition, the *code official* shall order the necessary work to be done, including the boarding up of openings, to render such *structure* temporarily safe whether or not the legal procedure herein described has been instituted; and shall cause such other action to be taken as the *code official* deems necessary to meet such emergency.

112.3 Closing streets. When necessary for public safety, the *code official* shall temporarily close structures and close, or order the authority having jurisdiction to close, sidewalks, streets, *public ways* and places adjacent to unsafe structures, and prohibit the same from being utilized.

112.4 Emergency repairs. For the purposes of this section, the *code official* shall employ the necessary labor and materials to perform the required work as expeditiously as possible.

112.5 Costs of emergency repairs. Costs incurred in the performance of emergency work shall be paid by the jurisdiction. The legal counsel of the jurisdiction shall institute appropriate action against the *owner* of the *premises* or *owner*'s authorized agent where the unsafe *structure* is or was located for the recovery of such costs.

112.6 Hearing. Any person ordered to take emergency measures shall comply with such order forthwith. Any affected person shall thereafter, upon petition directed to the appeals board, be afforded a hearing as described in this code.

SECTION 113
DEMOLITION

113.1 General. The *code official* shall order the *owner* or *owner*'s authorized agent of any *premises* upon which is located any *structure*, which in the *code official's* or *owner*'s authorized agent judgment after review is so deteriorated or dilapidated or has become so out of repair as to be dangerous, unsafe, insanitary or otherwise unfit for human habitation or *occupancy*, and such that it is unreasonable to repair the *structure*, to demolish and remove such *structure*; or if such *structure* is capable of being made safe by repairs, to repair and make safe and sanitary, or to board up and hold for future repair or to demolish and remove at the *owner's* option; or where there has been a cessation of normal construction of any *structure* for a period of more than two years, the *code official* shall order the *owner* or owner's authorized agent to demolish and remove such *structure*, or board up until future repair. Boarding the building up for future repair shall not extend beyond one year, unless *approved* by the building official.

113.2 Notices and orders. Notices and orders shall comply with Section 111.4.

113.3 Failure to comply. If the *owner* of a *premises* or *owner*'s authorized agent fails to comply with a demolition order within the time prescribed, the *code official* shall cause the *structure* to be demolished and removed, either through an available public agency or by contract or arrangement with private persons, and the cost of such demolition and removal shall be charged against the real estate upon which the *structure* is located and shall be a lien upon such real estate.

113.4 Salvage materials. Where any *structure* has been ordered demolished and removed, the governing body or other designated officer under said contract or arrangement aforesaid shall have the right to sell the salvage and valuable materials. The net proceeds of such sale, after deducting the expenses of such demolition and removal, shall be promptly remitted with a report of such sale or transaction, including the items of expense and the amounts deducted, for the person who is entitled thereto, subject to any order of a court. If such a surplus does not remain to be turned over, the report shall so state.

CHAPTER 2
DEFINITIONS

User note:

About this chapter: *Codes, by their very nature, are technical documents. Every word, term and punctuation mark can add to or change the meaning of a technical requirement. It is necessary to maintain a consensus on the specific meaning of each term contained in the code. Chapter 2 performs this function by stating clearly what specific terms mean for the purpose of the code.*

SECTION 201
GENERAL

201.1 Scope. Unless otherwise expressly stated, the following terms shall, for the purposes of this code, have the meanings shown in this chapter.

201.2 Interchangeability. Words stated in the present tense include the future; words stated in the masculine gender include the feminine and neuter; the singular number includes the plural and the plural, the singular.

201.3 Terms defined in other codes. Where terms are not defined in this code and are defined in the *International Building Code, International Existing Building Code, International Fire Code, International Fuel Gas Code, International Mechanical Code, International Plumbing Code, International Residential Code, International Zoning Code* or *NFPA 70,* such terms shall have the meanings ascribed to them as stated in those codes.

> **Exception:** When used within this code, the terms unsafe and dangerous shall have only the meanings ascribed to them in this code and shall not have the meanings ascribed to them by the International Existing Building Code.

201.4 Terms not defined. Where terms are not defined through the methods authorized by this section, such terms shall have ordinarily accepted meanings such as the context implies.

201.5 Parts. Whenever the words "*dwelling unit,*" "dwelling," "*premises,*" "building," "*rooming house,*" "*rooming unit,*" "*housekeeping unit*" or "story" are stated in this code, they shall be construed as though they were followed by the words "or any part thereof."

SECTION 202
GENERAL DEFINITIONS

ANCHORED. Secured in a manner that provides positive connection.

[A] APPROVED. Acceptable to the *code official.*

BASEMENT. That portion of a building that is partly or completely below grade.

BATHROOM. A room containing plumbing fixtures including a bathtub or shower.

BEDROOM. Any room or space used or intended to be used for sleeping purposes in either a dwelling or *sleeping unit.*

[A] CODE OFFICIAL. The official who is charged with the administration and enforcement of this code, or any duly authorized representative.

CONDEMN. To adjudge unfit for *occupancy.*

COST OF SUCH DEMOLITION OR EMERGENCY REPAIRS. The costs shall include the actual costs of the demolition or repair of the *structure* less revenues obtained if salvage was conducted prior to demolition or repair. Costs shall include, but not be limited to, expenses incurred or necessitated related to demolition or emergency repairs, such as asbestos survey and abatement if necessary; costs of inspectors, testing agencies or experts retained relative to the demolition or emergency repairs; costs of testing; surveys for other materials that are controlled or regulated from being dumped in a landfill; title searches; mailing(s); postings; recording; and attorney fees expended for recovering of the cost of emergency repairs or to obtain or enforce an order of demolition made by a *code official,* the governing body or board of appeals.

DETACHED. When a structural element is physically disconnected from another and that connection is necessary to provide a positive connection.

DETERIORATION. To weaken, disintegrate, corrode, rust or decay and lose effectiveness.

[A] DWELLING UNIT. A single unit providing complete, independent living facilities for one or more persons, including permanent provisions for living, sleeping, eating, cooking and sanitation.

[Z] EASEMENT. That portion of land or property reserved for present or future use by a person or agency other than the legal fee *owner*(s) of the property. The *easement* shall be permitted to be for use under, on or above said lot or lots.

EMERGENCY ESCAPE AND RESCUE OPENING. An operable exterior window, door or other similar device that provides for a means of escape and access for rescue in the event of an emergency.

EQUIPMENT SUPPORT. Those structural members or assemblies of members or manufactured elements, including braces, frames, lugs, snuggers, hangers or saddles, that transmit gravity load, lateral load and operating load between the equipment and the *structure.*

EXTERIOR PROPERTY. The open space on the *premises* and on adjoining property under the control of *owners* or *operators* of such *premises.*

GARBAGE. The animal or vegetable waste resulting from the handling, preparation, cooking and consumption of food.

[BE] GUARD. A building component or a system of building components located at or near the open sides of elevated walking surfaces that minimizes the possibility of a fall from the walking surface to a lower level.

[BG] HABITABLE SPACE. Space in a *structure* for living, sleeping, eating or cooking. *Bathrooms, toilet rooms*, closets, halls, storage or utility spaces, and similar areas are not considered *habitable spaces.*

[A] HISTORIC BUILDING. Any building or *structure* that is one or more of the following:

1. Listed or certified as eligible for listing, by the State Historic Preservation Officer or the Keeper of the National Register of Historic Places, in the National Register of Historic Places.
2. Designated as historic under an applicable state or local law.
3. Certified as a contributing resource within a National Register or state or locally designated historic district.

HOUSEKEEPING UNIT. A room or group of rooms forming a single *habitable space* equipped and intended to be used for living, sleeping, cooking and eating that does not contain, within such a unit, a toilet, lavatory and bathtub or shower.

IMMINENT DANGER. A condition that could cause serious or life-threatening injury or death at any time.

INFESTATION. The presence, within or contiguous to, a *structure* or *premises* of insects, rodents, vermin or other pests.

INOPERABLE MOTOR VEHICLE. A vehicle that cannot be driven upon the public streets for reason including but not limited to being unlicensed, wrecked, abandoned, in a state of disrepair, or incapable of being moved under its own power.

[A] LABELED. Equipment, materials or products to which have been affixed a label, seal, symbol or other identifying mark of a nationally recognized testing laboratory, *approved* agency or other organization concerned with product evaluation that maintains periodic inspection of the production of the above-*labeled* items and whose labeling indicates either that the equipment, material or product meets identified standards or has been tested and found suitable for a specified purpose.

LET FOR OCCUPANCY or LET. To permit, provide or offer possession or *occupancy* of a dwelling, *dwelling unit, rooming unit*, building, premise or *structure* by a person who is or is not the legal *owner* of record thereof, pursuant to a written or unwritten lease, agreement or license, or pursuant to a recorded or unrecorded agreement of contract for the sale of land.

NEGLECT. The lack of proper maintenance for a building or *structure.*

[A] OCCUPANCY. The purpose for which a building or portion thereof is utilized or occupied.

OCCUPANT. Any individual living or sleeping in a building, or having possession of a space within a building.

OPENABLE AREA. That part of a window, skylight or door which is available for unobstructed *ventilation* and which opens directly to the outdoors.

OPERATOR. Any person who has charge, care or control of a *structure* or *premises* that is let or offered for *occupancy.*

[A] OWNER. Any person, agent, *operator*, firm or corporation having legal or equitable interest in the property; or recorded in the official records of the state, county or municipality as holding title to the property; or otherwise having control of the property, including the guardian of the estate of any such person, and the executor or administrator of the estate of such person if ordered to take possession of real property by a court.

[A] PERSON. An individual, corporation, partnership or any other group acting as a unit.

PEST ELIMINATION. The control and elimination of insects, rodents or other pests by eliminating their harborage places; by removing or making inaccessible materials that serve as their food or water; by other *approved pest elimination* methods.

[A] PREMISES. A lot, plot or parcel of land, *easement* or *public way*, including any structures thereon.

[A] PUBLIC WAY. Any street, alley or other parcel of land that: is open to the outside air; leads to a street; has been deeded, dedicated or otherwise permanently appropriated to the public for public use; and has a clear width and height of not less than 10 feet (3048 mm).

ROOMING HOUSE. A building arranged or occupied for lodging, with or without meals, for compensation and not occupied as a one- or two-family dwelling.

ROOMING UNIT. Any room or group of rooms forming a single habitable unit occupied or intended to be occupied for sleeping or living, but not for cooking purposes.

RUBBISH. Combustible and noncombustible waste materials, except garbage; the term shall include the residue from the burning of wood, coal, coke and other combustible materials, paper, rags, cartons, boxes, wood, excelsior, rubber, leather, tree branches, *yard* trimmings, tin cans, metals, mineral matter, glass, crockery and dust and other similar materials.

[A] SLEEPING UNIT. A room or space in which people sleep, which can also include permanent provisions for living, eating and either sanitation or kitchen facilities, but not both. Such rooms and spaces that are also part of a *dwelling unit* are not *sleeping units.*

STRICT LIABILITY OFFENSE. An offense in which the prosecution in a legal proceeding is not required to prove criminal intent as a part of its case. It is enough to prove that the defendant either did an act which was prohibited, or failed to do an act which the defendant was legally required to do.

[A] STRUCTURE. That which is built or constructed.

TENANT. A person, corporation, partnership or group, whether or not the legal *owner* of record, occupying a building or portion thereof as a unit.

TOILET ROOM. A room containing a water closet or urinal but not a bathtub or shower.

ULTIMATE DEFORMATION. The deformation at which failure occurs and that shall be deemed to occur if the sustainable load reduces to 80 percent or less of the maximum strength.

[M] VENTILATION. The natural or mechanical process of supplying conditioned or unconditioned air to, or removing such air from, any space.

WORKMANLIKE. Executed in a skilled manner; e.g., generally plumb, level, square, in line, undamaged and without marring adjacent work.

[Z] YARD. An open space on the same lot with a *structure*.

CHAPTER 3
GENERAL REQUIREMENTS

User note:

About this chapter: *Chapter 3 is broad in scope and includes a variety of requirements for the maintenance of exterior property areas, as well as the interior and exterior elements of the structure, that are intended to maintain a minimum level of safety and sanitation for both the general public and the occupants of a structure, and to maintain a building's structural and weather-resistance performance. Specifically, Chapter 3 contains criteria for the maintenance of building components; vacant structures and land; the safety, sanitation and appearance of the interior and exterior of structures and all exterior property areas; accessory structures; extermination of insects and rodents; access barriers to swimming pools, spas and hot tubs; vehicle storage and owner/occupant responsibilities.*

SECTION 301
GENERAL

301.1 Scope. The provisions of this chapter shall govern the minimum conditions and the responsibilities of persons for maintenance of structures, equipment and *exterior property.*

301.2 Responsibility. The *owner* of the *premises* shall maintain the structures and *exterior property* in compliance with these requirements, except as otherwise provided for in this code. A person shall not occupy as *owner-occupant* or permit another person to occupy *premises* that are not in a sanitary and safe condition and that do not comply with the requirements of this chapter. *Occupants* of a *dwelling unit, rooming unit* or *housekeeping unit* are responsible for keeping in a clean, sanitary and safe condition that part of the *dwelling unit, rooming unit, housekeeping unit* or *premises* they occupy and control.

301.3 Vacant structures and land. Vacant structures and *premises* thereof or vacant land shall be maintained in a clean, safe, secure and sanitary condition as provided herein so as not to cause a blighting problem or adversely affect the public health or safety.

SECTION 302
EXTERIOR PROPERTY AREAS

302.1 Sanitation. *Exterior property* and *premises* shall be maintained in a clean, safe and sanitary condition. The *occupant* shall keep that part of the *exterior property* that such *occupant* occupies or controls in a clean and sanitary condition.

302.2 Grading and drainage. *Premises* shall be graded and maintained to prevent the erosion of soil and to prevent the accumulation of stagnant water thereon, or within any *structure* located thereon.

Exception: *Approved* retention areas and reservoirs.

302.3 Sidewalks and driveways. Sidewalks, walkways, stairs, driveways, parking spaces and similar areas shall be kept in a proper state of repair, and maintained free from hazardous conditions.

302.4 Weeds. *Premises* and *exterior property* shall be maintained free from weeds or plant growth in excess of **[JURISDICTION TO INSERT HEIGHT IN INCHES]**. Noxious weeds shall be prohibited. Weeds shall be defined as all grasses, annual plants and vegetation, other than trees or shrubs provided; however, this term shall not include cultivated flowers and gardens.

Upon failure of the *owner* or agent having charge of a property to cut and destroy weeds after service of a notice of violation, they shall be subject to prosecution in accordance with Section 108.3 and as prescribed by the authority having jurisdiction. Upon failure to comply with the notice of violation, any duly authorized employee of the jurisdiction or contractor hired by the jurisdiction shall be authorized to enter upon the property in violation and cut and destroy the weeds growing thereon, and the costs of such removal shall be paid by the *owner* or agent responsible for the property.

302.5 Rodent harborage. Structures and *exterior property* shall be kept free from rodent harborage and *infestation.* Where rodents are found, they shall be promptly exterminated by *approved* processes that will not be injurious to human health. After *pest elimination,* proper precautions shall be taken to eliminate rodent harborage and prevent reinfestation.

302.6 Exhaust vents. Pipes, ducts, conductors, fans or blowers shall not discharge gases, steam, vapor, hot air, grease, smoke, odors or other gaseous or particulate wastes directly on abutting or adjacent public or private property or that of another *tenant.*

302.7 Accessory structures. Accessory structures, including detached garages, fences and walls, shall be maintained structurally sound and in good repair.

302.8 Motor vehicles. Except as provided for in other regulations, inoperative or unlicensed motor vehicles shall not be parked, kept or stored on any *premises,* and vehicles shall not at any time be in a state of major disassembly, disrepair, or in the process of being stripped or dismantled. Painting of vehicles is prohibited unless conducted inside an *approved* spray booth.

Exception: A vehicle of any type is permitted to undergo major overhaul, including body work, provided that such work is performed inside a *structure* or similarly enclosed area designed and *approved* for such purposes.

302.9 Defacement of property. A person shall not willfully or wantonly damage, mutilate or deface any exterior surface of any *structure* or building on any private or public property by placing thereon any marking, carving or graffiti.

It shall be the responsibility of the *owner* to restore said surface to an *approved* state of maintenance and repair.

SECTION 303 SWIMMING POOLS, SPAS AND HOT TUBS

303.1 Swimming pools. Swimming pools shall be maintained in a clean and sanitary condition, and in good repair.

303.2 Enclosures. Private swimming pools, hot tubs and spas, containing water more than 24 inches (610 mm) in depth shall be completely surrounded by a fence or barrier not less than 48 inches (1219 mm) in height above the finished ground level measured on the side of the barrier away from the pool. Gates and doors in such barriers shall be self-closing and self-latching. Where the self-latching device is less than 54 inches (1372 mm) above the bottom of the gate, the release mechanism shall be located on the pool side of the gate. Self-closing and self-latching gates shall be maintained such that the gate will positively close and latch when released from an open position of 6 inches (152 mm) from the gatepost. An existing pool enclosure shall not be removed, replaced or changed in a manner that reduces its effectiveness as a safety barrier.

Exception: Spas or hot tubs with a safety cover that complies with ASTM F1346 shall be exempt from the provisions of this section.

SECTION 304 EXTERIOR STRUCTURE

304.1 General. The exterior of a *structure* shall be maintained in good repair, structurally sound and sanitary so as not to pose a threat to the public health, safety or welfare.

304.1.1 Unsafe conditions. The following conditions shall be determined as unsafe and shall be repaired or replaced to comply with the *International Building Code* or the *International Existing Building Code* as required for existing buildings:

1. The nominal strength of any structural member is exceeded by nominal loads, the load effects or the required strength.
2. The *anchorage* of the floor or roof to walls or columns, and of walls and columns to foundations is not capable of resisting all nominal loads or load effects.
3. Structures or components thereof that have reached their limit state.
4. Siding and masonry joints including joints between the building envelope and the perimeter of windows, doors and skylights are not maintained, weather resistant or water tight.
5. Structural members that have evidence of *deterioration* or that are not capable of safely supporting all nominal loads and load effects.
6. Foundation systems that are not firmly supported by footings, are not plumb and free from open cracks and breaks, are not properly *anchored* or are not capable of supporting all nominal loads and resisting all load effects.
7. Exterior walls that are not *anchored* to supporting and supported elements or are not plumb and free of holes, cracks or breaks and loose or rotting materials, are not properly *anchored* or are not capable of supporting all nominal loads and resisting all load effects.
8. Roofing or roofing components that have defects that admit rain, roof surfaces with inadequate drainage, or any portion of the roof framing that is not in good repair with signs of *deterioration*, fatigue or without proper anchorage and incapable of supporting all nominal loads and resisting all load effects.
9. Flooring and flooring components with defects that affect serviceability or flooring components that show signs of *deterioration* or fatigue, are not properly *anchored* or are incapable of supporting all nominal loads and resisting all load effects.
10. Veneer, cornices, belt courses, corbels, trim, wall facings and similar decorative features not properly *anchored* or that are *anchored* with connections not capable of supporting all nominal loads and resisting all load effects.
11. Overhang extensions or projections including, but not limited to, trash chutes, canopies, marquees, signs, awnings, fire escapes, standpipes and exhaust ducts not properly *anchored* or that are *anchored* with connections not capable of supporting all nominal loads and resisting all load effects.
12. Exterior stairs, decks, porches, balconies and all similar appurtenances attached thereto, including *guards* and handrails, are not structurally sound, not properly *anchored* or that are *anchored* with connections not capable of supporting all nominal loads and resisting all load effects.
13. Chimneys, cooling towers, smokestacks and similar appurtenances not structurally sound or not properly *anchored*, or that are *anchored* with connections not capable of supporting all nominal loads and resisting all load effects.

Exceptions:

1. Where substantiated otherwise by an *approved* method.
2. Demolition of unsafe conditions shall be permitted where *approved* by the *code official*.

304.2 Protective treatment. Exterior surfaces, including but not limited to, doors, door and window frames, cornices, porches, trim, balconies, decks and fences, shall be maintained in good condition. Exterior wood surfaces, other than decay-resistant woods, shall be protected from the elements and decay by painting or other protective covering or treatment. Peeling, flaking and chipped paint shall be eliminated and surfaces repainted. Siding and masonry joints, as well as those between the building envelope and the perimeter of windows, doors and skylights, shall be maintained weather resistant and water tight. Metal surfaces subject to rust or corrosion shall be coated to inhibit such rust and corrosion, and surfaces with rust or corrosion shall be stabilized and coated to inhibit future rust and corrosion. Oxidation stains shall be removed from exterior surfaces. Surfaces designed for stabilization by oxidation are exempt from this requirement.

[F] 304.3 Premises identification. Buildings shall have *approved* address numbers placed in a position to be plainly legible and visible from the street or road fronting the property. These numbers shall contrast with their background. Address numbers shall be Arabic numerals or alphabet letters. Numbers shall be not less than 4 inches (102 mm) in height with a minimum stroke width of 0.5 inch (12.7 mm).

304.4 Structural members. Structural members shall be maintained free from *deterioration*, and shall be capable of safely supporting the imposed dead and live loads.

304.5 Foundation walls. Foundation walls shall be maintained plumb and free from open cracks and breaks and shall be kept in such condition so as to prevent the entry of rodents and other pests.

304.6 Exterior walls. Exterior walls shall be free from holes, breaks, and loose or rotting materials; and maintained weatherproof and properly surface coated where required to prevent *deterioration*.

304.7 Roofs and drainage. The roof and flashing shall be sound, tight and not have defects that admit rain. Roof drainage shall be adequate to prevent dampness or *deterioration* in the walls or interior portion of the *structure*. Roof drains, gutters and downspouts shall be maintained in good repair and free from obstructions. Roof water shall not be discharged in a manner that creates a public nuisance.

304.8 Decorative features. Cornices, belt courses, corbels, terra cotta trim, wall facings and similar decorative features shall be maintained in good repair with proper *anchorage* and in a safe condition.

304.9 Overhang extensions. Overhang extensions including, but not limited to, canopies, marquees, signs, metal awnings, fire escapes, standpipes and exhaust ducts shall be maintained in good repair and be properly *anchored* so as to be kept in a sound condition. Where required, all exposed surfaces of metal or wood shall be protected from the elements and against decay or rust by periodic application of weather-coating materials, such as paint or similar surface treatment.

304.10 Stairways, decks, porches and balconies. Every exterior stairway, deck, porch and balcony, and all appurtenances attached thereto, shall be maintained structurally sound, in good repair, with proper *anchorage* and capable of supporting the imposed loads.

304.11 Chimneys and towers. Chimneys, cooling towers, smoke stacks, and similar appurtenances shall be maintained structurally safe and sound, and in good repair. Exposed surfaces of metal or wood shall be protected from the elements and against decay or rust by periodic application of weather-coating materials, such as paint or similar surface treatment.

304.12 Handrails and guards. Every handrail and *guard* shall be firmly fastened and capable of supporting normally imposed loads and shall be maintained in good condition.

304.13 Window, skylight and door frames. Every window, skylight, door and frame shall be kept in sound condition, good repair and weather tight.

304.13.1 Glazing. Glazing materials shall be maintained free from cracks and holes.

304.13.2 Openable windows. Every window, other than a fixed window, shall be easily openable and capable of being held in position by window hardware.

304.14 Insect screens. During the period from **[DATE]** to **[DATE]**, every door, window and other outside opening required for *ventilation* of habitable rooms, food preparation areas, food service areas or any areas where products to be included or utilized in food for human consumption are processed, manufactured, packaged or stored shall be supplied with *approved* tightly fitting screens of minimum 16 mesh per inch (16 mesh per 25 mm), and every screen door used for insect control shall have a self-closing device in good working condition.

Exception: Screens shall not be required where other *approved* means, such as air curtains or insect repellent fans, are employed.

304.15 Doors. Exterior doors, door assemblies, operator systems if provided, and hardware shall be maintained in good condition. Locks at all entrances to *dwelling units* and sleeping units shall tightly secure the door. Locks on means of egress doors shall be in accordance with Section 702.3.

304.16 Basement hatchways. Every *basement* hatchway shall be maintained to prevent the entrance of rodents, rain and surface drainage water.

304.17 Guards for basement windows. Every *basement* window that is openable shall be supplied with rodent shields, storm windows or other *approved* protection against the entry of rodents.

304.18 Building security. Doors, windows or hatchways for *dwelling units*, room units or *housekeeping units* shall be provided with devices designed to provide security for the *occupants* and property within.

304.18.1 Doors. Doors providing access to a *dwelling unit, rooming unit* or *housekeeping unit* that is rented, leased or let shall be equipped with a deadbolt lock designed to be readily openable from the side from which egress is to be made without the need for keys, special knowledge or effort and shall have a minimum lock throw of 1 inch (25 mm). Such deadbolt locks shall be installed according to the manufacturer's specifications and maintained in good working order. For the purpose of this section, a sliding bolt shall not be considered an acceptable deadbolt lock.

304.18.2 Windows. Operable windows located in whole or in part within 6 feet (1828 mm) above ground level or a walking surface below that provide access to a *dwelling unit, rooming unit* or *housekeeping unit* that is rented, leased or let shall be equipped with a window sash locking device.

304.18.3 Basement hatchways. *Basement* hatchways that provide access to a *dwelling unit, rooming unit* or *housekeeping unit* that is rented, leased or let shall be equipped with devices that secure the units from unauthorized entry.

304.19 Gates. Exterior gates, gate assemblies, operator systems if provided, and hardware shall be maintained in good condition. Latches at all entrances shall tightly secure the gates.

SECTION 305
INTERIOR STRUCTURE

305.1 General. The interior of a *structure* and equipment therein shall be maintained in good repair, structurally sound and in a sanitary condition. *Occupants* shall keep that part of the *structure* that they occupy or control in a clean and sanitary condition. Every *owner* of a *structure* containing a *rooming house, housekeeping units*, a hotel, a dormitory, two or more *dwelling units* or two or more nonresidential *occupancies*, shall maintain, in a clean and sanitary condition, the shared or public areas of the *structure* and *exterior property*.

305.1.1 Unsafe conditions. The following conditions shall be determined as unsafe and shall be repaired or replaced to comply with the *International Building Code* or the *International Existing Building Code* as required for existing buildings:

1. The nominal strength of any structural member is exceeded by nominal loads, the load effects or the required strength.
2. The anchorage of the floor or roof to walls or columns, and of walls and columns to foundations is not capable of resisting all nominal loads or load effects.
3. Structures or components thereof that have reached their limit state.
4. Structural members are incapable of supporting nominal loads and load effects.
5. Stairs, landings, balconies and all similar walking surfaces, including *guards* and handrails, are not structurally sound, not properly *anchored* or are *anchored* with connections not capable of supporting all nominal loads and resisting all load effects.
6. Foundation systems that are not firmly supported by footings are not plumb and free from open cracks and breaks, are not properly *anchored* or are not capable of supporting all nominal loads and resisting all load effects.

Exceptions:

1. Where substantiated otherwise by an *approved* method.
2. Demolition of unsafe conditions shall be permitted where *approved* by the *code official.*

305.2 Structural members. Structural members shall be maintained structurally sound, and be capable of supporting the imposed loads.

305.3 Interior surfaces. Interior surfaces, including windows and doors, shall be maintained in good, clean and sanitary condition. Peeling, chipping, flaking or abraded paint shall be repaired, removed or covered. Cracked or loose plaster, decayed wood and other defective surface conditions shall be corrected.

305.4 Stairs and walking surfaces. Every stair, ramp, landing, balcony, porch, deck or other walking surface shall be maintained in sound condition and good repair.

305.5 Handrails and guards. Every handrail and *guard* shall be firmly fastened and capable of supporting normally imposed loads and shall be maintained in good condition.

305.6 Interior doors. Every interior door shall fit reasonably well within its frame and shall be capable of being opened and closed by being properly and securely attached to jambs, headers or tracks as intended by the manufacturer of the attachment hardware.

SECTION 306
COMPONENT SERVICEABILITY

306.1 General. The components of a *structure* and equipment therein shall be maintained in good repair, structurally sound and in a sanitary condition.

306.1.1 Unsafe conditions. Where any of the following conditions cause the component or system to be beyond its limit state, the component or system shall be determined as unsafe and shall be repaired or replaced to comply with the *International Building Code* or the *International Existing Building Code* as required for existing buildings:

1. Soils that have been subjected to any of the following conditions:
    1.1. Collapse of footing or foundation system.
    1.2. Damage to footing, foundation, concrete or other structural element due to soil expansion.
    1.3. Adverse effects to the design strength of footing, foundation, concrete or other structural element due to a chemical reaction from the soil.
    1.4. Inadequate soil as determined by a geotechnical investigation.
    1.5. Where the allowable bearing capacity of the soil is in doubt.

1.6. Adverse effects to the footing, foundation, concrete or other structural element due to the ground water table.

2. Concrete that has been subjected to any of the following conditions:
   2.1. *Deterioration.*
   2.2. *Ultimate deformation.*
   2.3. Fractures.
   2.4. Fissures.
   2.5. Spalling.
   2.6. Exposed reinforcement.
   2.7. *Detached,* dislodged or failing connections.
3. Aluminum that has been subjected to any of the following conditions:
   3.1. *Deterioration.*
   3.2. Corrosion.
   3.3. Elastic deformation.
   3.4. *Ultimate deformation.*
   3.5. Stress or strain cracks.
   3.6. Joint fatigue.
   3.7. *Detached,* dislodged or failing connections.
4. Masonry that has been subjected to any of the following conditions:
   4.1. *Deterioration.*
   4.2. *Ultimate deformation.*
   4.3. Fractures in masonry or mortar joints.
   4.4. Fissures in masonry or mortar joints.
   4.5. Spalling.
   4.6. Exposed reinforcement.
   4.7. *Detached,* dislodged or failing connections.
5. Steel that has been subjected to any of the following conditions:
   5.1. *Deterioration.*
   5.2. Elastic deformation.
   5.3. *Ultimate deformation.*
   5.4. Metal fatigue.
   5.5. *Detached,* dislodged or failing connections.
6. Wood that has been subjected to any of the following conditions:
   6.1. Ultimate deformation.
   6.2. Deterioration.
   6.3. Damage from insects, rodents and other vermin.
   6.4. Fire damage beyond charring.
   6.5. Significant splits and checks.
   6.6. Horizontal shear cracks.
   6.7. Vertical shear cracks.
   6.8. Inadequate support.
   6.9. *Detached,* dislodged or failing connections.
   6.10. Excessive cutting and notching.

Exceptions:

1. Where substantiated otherwise by an *approved* method.
2. Demolition of unsafe conditions shall be permitted where *approved* by the *code official.*

SECTION 307
HANDRAILS AND GUARDRAILS

307.1 General. Every exterior and interior flight of stairs having more than four risers shall have a handrail on one side of the stair and every open portion of a stair, landing, balcony, porch, deck, ramp or other walking surface that is more than 30 inches (762 mm) above the floor or grade below shall have *guards.* Handrails shall be not less than 30 inches (762 mm) in height or more than 42 inches (1067 mm) in height measured vertically above the nosing of the tread or above the finished floor of the landing or walking surfaces. *Guards* shall be not less than 30 inches (762 mm) in height above the floor of the landing, balcony, porch, deck, or ramp or other walking surface.

Exception: *Guards* shall not be required where exempted by the adopted building code.

SECTION 308
RUBBISH AND GARBAGE

308.1 Accumulation of rubbish or garbage. *Exterior property* and *premises,* and the interior of every *structure,* shall be free from any accumulation of *rubbish* or garbage.

308.2 Disposal of rubbish. Every *occupant* of a *structure* shall dispose of all *rubbish* in a clean and sanitary manner by placing such *rubbish* in *approved* containers.

308.2.1 Rubbish storage facilities. The *owner* of every occupied *premises* shall supply *approved* covered containers for *rubbish,* and the *owner* of the *premises* shall be responsible for the removal of *rubbish.*

308.2.2 Refrigerators. Refrigerators and similar equipment not in operation shall not be discarded, abandoned or stored on *premises* without first removing the doors.

308.3 Disposal of garbage. Every *occupant* of a *structure* shall dispose of garbage in a clean and sanitary manner by placing such garbage in an *approved* garbage disposal facility or *approved* garbage containers.

308.3.1 Garbage facilities. The *owner* of every dwelling shall supply one of the following: an *approved* mechanical food waste grinder in each *dwelling unit;* an *approved* incinerator unit in the *structure* available to the *occupants* in each *dwelling unit;* or an *approved* leakproof, covered, outside garbage container.

308.3.2 Containers. The *operator* of every establishment producing garbage shall provide, and at all times cause to be utilized, *approved* leakproof containers provided with close-fitting covers for the storage of such materials until removed from the *premises* for disposal.

SECTION 309
PEST ELIMINATION

309.1 Infestation. Structures shall be kept free from insect and rodent *infestation*. Structures in which insects or rodents are found shall be promptly exterminated by *approved* processes that will not be injurious to human health. After *pest elimination*, proper precautions shall be taken to prevent reinfestation.

309.2 Owner. The *owner* of any *structure* shall be responsible for *pest elimination* within the *structure* prior to renting or leasing the *structure*.

309.3 Single occupant. The *occupant* of a one-family dwelling or of a single-*tenant* nonresidential *structure* shall be responsible for *pest elimination* on the *premises*.

309.4 Multiple occupancy. The *owner* of a *structure* containing two or more *dwelling units*, a multiple *occupancy*, a *rooming house* or a nonresidential *structure* shall be responsible for *pest elimination* in the public or shared areas of the *structure* and *exterior property*. If *infestation* is caused by failure of an *occupant* to prevent such *infestation* in the area occupied, the *occupant* and *owner* shall be responsible for *pest elimination*.

309.5 Occupant. The *occupant* of any *structure* shall be responsible for the continued rodent and pest-free condition of the *structure*.

> **Exception:** Where the *infestations* are caused by defects in the *structure*, the *owner* shall be responsible for *pest elimination*.

CHAPTER 4

LIGHT, VENTILATION AND OCCUPANCY LIMITATIONS

User note:

About this chapter: *Chapter 4 sets forth requirements to establish the minimum environment for occupiable and habitable buildings by establishing the minimum criteria for light and ventilation and identifying occupancy limitations including minimum room width and area, minimum ceiling height and restrictions to prevent overcrowding.*

SECTION 401
GENERAL

401.1 Scope. The provisions of this chapter shall govern the minimum conditions and standards for light, *ventilation* and space for occupying a *structure*.

401.2 Responsibility. The *owner* of the *structure* shall provide and maintain light, *ventilation* and space conditions in compliance with these requirements. A person shall not occupy as *owner-occupant*, or permit another person to occupy, any *premises* that do not comply with the requirements of this chapter.

401.3 Alternative devices. In lieu of the means for natural light and *ventilation* herein prescribed, artificial light or mechanical *ventilation* complying with the *International Building Code* shall be permitted.

SECTION 402
LIGHT

402.1 Habitable spaces. Every *habitable space* shall have not less than one window of *approved* size facing directly to the outdoors or to a court. The minimum total glazed area for every *habitable space* shall be 8 percent of the floor area of such room. Wherever walls or other portions of a *structure* face a window of any room and such obstructions are located less than 3 feet (914 mm) from the window and extend to a level above that of the ceiling of the room, such window shall not be deemed to face directly to the outdoors nor to a court and shall not be included as contributing to the required minimum total window area for the room.

Exception: Where natural light for rooms or spaces without exterior glazing areas is provided through an adjoining room, the unobstructed opening to the adjoining room shall be not less than 8 percent of the floor area of the interior room or space, or not less than 25 square feet (2.33 m^2), whichever is greater. The exterior glazing area shall be based on the total floor area being served.

402.2 Common halls and stairways. Every common hall and stairway in residential *occupancies*, other than in one- and two-family dwellings, shall be lighted at all times with not less than a 60-watt standard incandescent light bulb for each 200 square feet (19 m^2) of floor area or equivalent illumination, provided that the spacing between lights shall not be greater than 30 feet (9144 mm). In other than residential occupancies, interior and exterior means of egress, stairways shall be illuminated at all times the building space served by the means of egress is occupied with not less than 1 footcandle (11 lux) at floors, landings and treads.

402.3 Other spaces. Other spaces shall be provided with natural or artificial light sufficient to permit the maintenance of sanitary conditions, and the safe *occupancy* of the space and utilization of the appliances, equipment and fixtures.

SECTION 403
VENTILATION

403.1 Habitable spaces. Every *habitable space* shall have not less than one openable window. The total openable area of the window in every room shall be equal to not less than 45 percent of the minimum glazed area required in Section 402.1.

Exception: Where rooms and spaces without openings to the outdoors are ventilated through an adjoining room, the unobstructed opening to the adjoining room shall be not less than 8 percent of the floor area of the interior room or space, but not less than 25 square feet (2.33 m^2). The *ventilation* openings to the outdoors shall be based on a total floor area being ventilated.

403.2 Bathrooms and toilet rooms. Every *bathroom* and *toilet room* shall comply with the *ventilation* requirements for *habitable spaces* as required by Section 403.1, except that a window shall not be required in such spaces equipped with a mechanical *ventilation* system. Air exhausted by a mechanical *ventilation* system from a *bathroom* or *toilet room* shall discharge to the outdoors and shall not be recirculated.

403.3 Cooking facilities. Unless *approved* through the certificate of *occupancy*, cooking shall not be permitted in any *rooming unit* or dormitory unit, and a cooking facility or appliance shall not be permitted to be present in the *rooming unit* or dormitory unit.

Exceptions:

1. Where specifically *approved* in writing by the *code official*.
2. Devices such as coffee pots and microwave ovens shall not be considered cooking appliances.

403.4 Process ventilation. Where injurious, toxic, irritating or noxious fumes, gases, dusts or mists are generated, a local exhaust *ventilation* system shall be provided to remove the contaminating agent at the source. Air shall be exhausted to the exterior and not be recirculated to any space.

403.5 Clothes dryer exhaust. Clothes dryer exhaust systems shall be independent of all other systems and shall be exhausted outside the *structure* in accordance with the manufacturer's instructions.

Exception: Listed and *labeled* condensing (ductless) clothes dryers.

SECTION 404 OCCUPANCY LIMITATIONS

404.1 Privacy. *Dwelling units*, hotel units, *housekeeping units, rooming units* and dormitory units shall be arranged to provide privacy and be separate from other adjoining spaces.

404.2 Minimum room widths. A habitable room, other than a kitchen, shall be not less than 7 feet (2134 mm) in any plan dimension. Kitchens shall have a minimum clear passageway of 3 feet (914 mm) between counterfronts and appliances or counterfronts and walls.

404.3 Minimum ceiling heights. *Habitable spaces*, hallways, corridors, laundry areas, *bathrooms, toilet rooms* and habitable *basement* areas shall have a minimum clear ceiling height of 7 feet (2134 mm).

Exceptions:

1. In one- and two-family dwellings, beams or girders spaced not less than 4 feet (1219 mm) on center and projecting not greater than 6 inches (152 mm) below the required ceiling height.
2. *Basement* rooms in one- and two-family dwellings occupied exclusively for laundry, study or recreation purposes, having a minimum ceiling height of 6 feet 8 inches (2033 mm) with a minimum clear height of 6 feet 4 inches (1932 mm) under beams, girders, ducts and similar obstructions.
3. Rooms occupied exclusively for sleeping, study or similar purposes and having a sloped ceiling over all or part of the room, with a minimum clear ceiling height of 7 feet (2134 mm) over not less than one-third of the required minimum floor area. In calculating the floor area of such rooms, only those portions of the floor area with a minimum clear ceiling height of 5 feet (1524 mm) shall be included.

404.4 Bedroom and living room requirements. Every *bedroom* and living room shall comply with the requirements of Sections 404.4.1 through 404.4.5.

404.4.1 Room area. Every living room shall contain not less than 120 square feet (11.2 m^2) and every bedroom shall contain not less than 70 square feet (6.5 m^2) and every bedroom occupied by more than one person shall contain not less than 50 square feet (4.6 m^2) of floor area for each *occupant* thereof.

404.4.2 Access from bedrooms. *Bedrooms* shall not constitute the only means of access to other *bedrooms* or *habitable spaces* and shall not serve as the only means of egress from other *habitable spaces.*

Exception: Units that contain fewer than two *bedrooms.*

404.4.3 Water closet accessibility. Every *bedroom* shall have access to not less than one water closet and one lavatory without passing through another *bedroom.* Every *bedroom* in a *dwelling unit* shall have access to not less than one water closet and lavatory located in the same story as the *bedroom* or an adjacent story.

404.4.4 Prohibited occupancy. Kitchens and nonhabitable spaces shall not be used for sleeping purposes.

404.4.5 Other requirements. *Bedrooms* shall comply with the applicable provisions of this code including, but not limited to, the light, *ventilation*, room area, ceiling height and room width requirements of this chapter; the plumbing facilities and water-heating facilities requirements of Chapter 5; the heating facilities and electrical receptacle requirements of Chapter 6; and the smoke detector and emergency escape requirements of Chapter 7.

404.5 Overcrowding. *Dwelling units* shall not be occupied by more *occupants* than permitted by the minimum area requirements of Table 404.5.

TABLE 404.5
MINIMUM AREA REQUIREMENTS

SPACE	MINIMUM AREA IN SQUARE FEET		
	1-2 occupants	3-5 occupants	6 or more occupants
Living room[a, b]	120	120	150
Dining room[a, b]	No requirement	80	100
Bedrooms	Shall comply with Section 404.4.1		

For SI: 1 square foot = 0.0929 m^2.

a. See Section 404.5.2 for combined living room/dining room spaces.

b. See Section 404.5.1 for limitations on determining the minimum occupancy area for sleeping purposes.

404.5.1 Sleeping area. The minimum occupancy area required by Table 404.5 shall not be included as a sleeping area in determining the minimum occupancy area for sleeping purposes. Sleeping areas shall comply with Section 404.4.

404.5.2 Combined spaces. Combined living room and dining room spaces shall comply with the requirements of Table 404.5 if the total area is equal to that required for separate rooms and if the space is located so as to function as a combination living room/dining room.

404.6 Efficiency unit. Nothing in this section shall prohibit an efficiency living unit from meeting the following requirements:

1. A unit occupied by not more than one *occupant* shall have a minimum clear floor area of 120 square feet (11.2 m^2). A unit occupied by not more than two *occupants* shall have a minimum clear floor area of 220 square feet (20.4 m^2). A unit occupied by three *occupants* shall have a minimum clear floor area of 320 square feet (29.7 m^2). These required areas shall be exclusive of the areas required by Items 2 and 3.
2. The unit shall be provided with a kitchen sink, cooking appliance and refrigeration facilities, each having a minimum clear working space of 30 inches

(762 mm) in front. Light and *ventilation* conforming to this code shall be provided.

3. The unit shall be provided with a separate *bathroom* containing a water closet, lavatory and bathtub or shower.
4. The maximum number of *occupants* shall be three.

404.7 Food preparation. Spaces to be occupied for food preparation purposes shall contain suitable space and equipment to store, prepare and serve foods in a sanitary manner. There shall be adequate facilities and services for the sanitary disposal of food wastes and refuse, including facilities for temporary storage.

CHAPTER 5

PLUMBING FACILITIES AND FIXTURE REQUIREMENTS

User note:

About this chapter: *Chapter 5 establishes minimum sanitary and clean conditions in occupied buildings by containing requirements for the installation, maintenance and location of plumbing systems and facilities, including the water supply system, water heating appliances, sewage disposal systems and related plumbing fixtures. Chapter 5 includes requirements for providing potable water to a building and the basic fixtures to effectively utilize and dispose of that water.*

SECTION 501 GENERAL

501.1 Scope. The provisions of this chapter shall govern the minimum plumbing systems, facilities and plumbing fixtures to be provided.

501.2 Responsibility. The *owner* of the *structure* shall provide and maintain such plumbing facilities and plumbing fixtures in compliance with these requirements. A person shall not occupy as *owner-occupant* or permit another person to occupy any *structure* or *premises* that does not comply with the requirements of this chapter.

SECTION 502 REQUIRED FACILITIES

[P] 502.1 Dwelling units. Every *dwelling unit* shall contain its own bathtub or shower, lavatory, water closet and kitchen sink that shall be maintained in a sanitary, safe working condition. The lavatory shall be placed in the same room as the water closet or located in close proximity to the door leading directly into the room in which such water closet is located. A kitchen sink shall not be used as a substitute for the required lavatory.

[P] 502.2 Rooming houses. Not less than one water closet, lavatory and bathtub or shower shall be supplied for each four *rooming units.*

[P] 502.3 Hotels. Where private water closets, lavatories and baths are not provided, one water closet, one lavatory and one bathtub or shower having access from a public hallway shall be provided for each 10 *occupants.*

[P] 502.4 Employees' facilities. Not less than one water closet, one lavatory and one drinking facility shall be available to employees.

[P] 502.4.1 Drinking facilities. Drinking facilities shall be a drinking fountain, water cooler, bottled water cooler or disposable cups next to a sink or water dispenser. Drinking facilities shall not be located in *toilet rooms* or *bathrooms.*

[P] 502.5 Public toilet facilities. Public toilet facilities shall be maintained in a safe, sanitary and working condition in accordance with the *International Plumbing Code.* Except for periodic maintenance or cleaning, public access and use shall be provided to the toilet facilities at all times during *occupancy* of the *premises.*

SECTION 503 TOILET ROOMS

[P] 503.1 Privacy. *Toilet rooms* and *bathrooms* shall provide privacy and shall not constitute the only passageway to a hall or other space, or to the exterior. A door and interior locking device shall be provided for all common or shared *bathrooms* and *toilet rooms* in a multiple dwelling.

[P] 503.2 Location. *Toilet rooms* and *bathrooms* serving hotel units, *rooming units* or dormitory units or *housekeeping units,* shall have access by traversing not more than one flight of stairs and shall have access from a common hall or passageway.

[P] 503.3 Location of employee toilet facilities. Toilet facilities shall have access from within the employees' working area. The required toilet facilities shall be located not more than one story above or below the employees' working area and the path of travel to such facilities shall not exceed a distance of 500 feet (152 m). Employee facilities shall either be separate facilities or combined employee and public facilities.

Exception: Facilities that are required for employees in storage structures or kiosks, which are located in adjacent structures under the same ownership, lease or control, shall not exceed a travel distance of 500 feet (152 m) from the employees' regular working area to the facilities.

[P] 503.4 Floor surface. In other than *dwelling units,* every *toilet room* floor shall be maintained to be a smooth, hard, nonabsorbent surface to permit such floor to be easily kept in a clean and sanitary condition.

SECTION 504 PLUMBING SYSTEMS AND FIXTURES

[P] 504.1 General. Plumbing fixtures shall be properly installed and maintained in working order, and shall be kept free from obstructions, leaks and defects and be capable of performing the function for which such plumbing fixtures are designed. Plumbing fixtures shall be maintained in a safe, sanitary and functional condition.

[P] 504.2 Fixture clearances. Plumbing fixtures shall have adequate clearances for usage and cleaning.

[P] 504.3 Plumbing system hazards. Where it is found that a plumbing system in a *structure* constitutes a hazard to the *occupants* or the *structure* by reason of inadequate service, inadequate venting, cross connection, backsiphonage, improper installation, *deterioration* or damage or for similar reasons, the *code official* shall require the defects to be corrected to eliminate the hazard.

SECTION 505
WATER SYSTEM

505.1 General. Every sink, lavatory, bathtub or shower, drinking fountain, water closet or other plumbing fixture shall be properly connected to either a public water system or to an *approved* private water system. Kitchen sinks, lavatories, laundry facilities, bathtubs and showers shall be supplied with hot or tempered and cold running water in accordance with the *International Plumbing Code.*

[P] 505.2 Contamination. The water supply shall be maintained free from contamination, and all water inlets for plumbing fixtures shall be located above the flood-level rim of the fixture. Shampoo basin faucets, janitor sink faucets and other hose bibs or faucets to which hoses are attached and left in place, shall be protected by an *approved* atmospheric-type vacuum breaker or an *approved* permanently attached hose connection vacuum breaker.

[P] 505.3 Supply. The water supply system shall be installed and maintained to provide a supply of water to plumbing fixtures, devices and appurtenances in sufficient volume and at pressures adequate to enable the fixtures to function properly, safely, and free from defects and leaks.

[P] 505.4 Water heating facilities. Water heating facilities shall be properly installed, maintained and capable of providing an adequate amount of water to be drawn at every required sink, lavatory, bathtub, shower and laundry facility at a temperature not less than 110°F (43°C). A gas-burning water heater shall not be located in any *bathroom, toilet room, bedroom* or other occupied room normally kept closed, unless adequate combustion air is provided. An *approved* combination temperature and pressure-relief valve and relief valve discharge pipe shall be properly installed and maintained on water heaters.

[P] 505.5 Nonpotable water reuse systems. Nonpotable water reuse systems and rainwater collection and conveyance systems shall be maintained in a safe and sanitary condition. Where such systems are not properly maintained, the systems shall be repaired to provide for safe and sanitary conditions, or the system shall be abandoned in accordance with Section 505.5.1.

[P] 505.5.1 Abandonment of systems. Where a nonpotable water reuse system or a rainwater collection and distribution system is not maintained or the *owner* ceases use of the system, the system shall be abandoned in accordance with Section 1301.10 of the *International Plumbing Code.*

SECTION 506
SANITARY DRAINAGE SYSTEM

[P] 506.1 General. Plumbing fixtures shall be properly connected to either a public sewer system or to an *approved* private sewage disposal system.

[P] 506.2 Maintenance. Every plumbing stack, vent, waste and sewer line shall function properly and be kept free from obstructions, leaks and defects.

[P] 506.3 Grease interceptors. Grease interceptors and automatic grease removal devices shall be maintained in accordance with this code and the manufacturer's installation instructions. Grease interceptors and automatic grease removal devices shall be regularly serviced and cleaned to prevent the discharge of oil, grease, and other substances harmful or hazardous to the building drainage system, the public sewer, the private sewage disposal system or the sewage treatment plant or processes. Records of maintenance, cleaning and repairs shall be available for inspection by the *code official.*

SECTION 507
STORM DRAINAGE

[P] 507.1 General. Drainage of roofs and paved areas, *yards* and courts, and other open areas on the *premises* shall not be discharged in a manner that creates a public nuisance.

CHAPTER 6
MECHANICAL AND ELECTRICAL REQUIREMENTS

User note:

About this chapter: *Chapter 6 establishes minimum performance requirements for heating, electrical and mechanical facilities serving existing structures, such as heating and air-conditioning equipment, appliances and their supporting systems; water heating equipment, appliances and systems; cooking equipment and appliances; ventilation and exhaust equipment; gas and liquid fuel distribution piping and components; fireplaces and solid fuel-burning appliances; chimneys and vents; electrical services; lighting fixtures; electrical receptacle outlets; electrical distribution system equipment, devices and wiring; and elevators, escalators and dumbwaiters.*

SECTION 601
GENERAL

601.1 Scope. The provisions of this chapter shall govern the minimum mechanical and electrical facilities and equipment to be provided.

601.2 Responsibility. The *owner* of the *structure* shall provide and maintain mechanical and electrical facilities and equipment in compliance with these requirements. A person shall not occupy as *owner-occupant* or permit another person to occupy any *premises* that does not comply with the requirements of this chapter.

SECTION 602
HEATING FACILITIES

602.1 Facilities required. Heating facilities shall be provided in structures as required by this section.

602.2 Residential occupancies. Dwellings shall be provided with heating facilities capable of maintaining a room temperature of 68°F (20°C) in all habitable rooms, *bathrooms* and *toilet rooms* based on the winter outdoor design temperature for the locality indicated in Appendix D of the *International Plumbing Code.* Cooking appliances shall not be used, nor shall portable unvented fuel-burning space heaters be used, as a means to provide required heating. The installation of one or more portable space heaters shall not be used to achieve compliance with this section.

Exception: In areas where the average monthly temperature is above 30°F (-1°C), a minimum temperature of 65°F (18°C) shall be maintained.

602.3 Heat supply. Every *owner* and *operator* of any building who rents, leases or lets one or more *dwelling units* or *sleeping units* on terms, either expressed or implied, to furnish heat to the *occupants* thereof shall supply heat during the period from **[DATE]** to **[DATE]** to maintain a minimum temperature of 68°F (20°C) in all habitable rooms, *bathrooms* and *toilet rooms.*

Exceptions:

1. When the outdoor temperature is below the winter outdoor design temperature for the locality, maintenance of the minimum room temperature shall not be required provided that the heating system is operating at its full design capacity. The winter outdoor design temperature for the locality shall be as indicated in Appendix D of the *International Plumbing Code.*
2. In areas where the average monthly temperature is above 30°F (-1°C), a minimum temperature of 65°F (18°C) shall be maintained.

602.4 Occupiable work spaces. Indoor occupiable work spaces shall be supplied with heat during the period from **[DATE]** to **[DATE]** to maintain a minimum temperature of 65°F (18°C) during the period the spaces are occupied.

Exceptions:

1. Processing, storage and operation areas that require cooling or special temperature conditions.
2. Areas in which persons are primarily engaged in vigorous physical activities.

602.5 Room temperature measurement. The required room temperatures shall be measured 3 feet (914 mm) above the floor near the center of the room and 2 feet (610 mm) inward from the center of each exterior wall.

SECTION 603
MECHANICAL EQUIPMENT

603.1 Mechanical equipment and appliances. Mechanical equipment, appliances, fireplaces, solid fuel-burning appliances, cooking appliances and water heating appliances shall be properly installed and maintained in a safe working condition, and shall be capable of performing the intended function.

603.2 Removal of combustion products. Fuel-burning equipment and appliances shall be connected to an *approved* chimney or vent.

Exception: Fuel-burning equipment and appliances that are *labeled* for unvented operation.

603.3 Clearances. Required clearances to combustible materials shall be maintained.

603.4 Safety controls. Safety controls for fuel-burning equipment shall be maintained in effective operation.

603.5 Combustion air. A supply of air for complete combustion of the fuel and for *ventilation* of the space containing the fuel-burning equipment shall be provided for the fuel-burning equipment.

603.6 Energy conservation devices. Devices intended to reduce fuel consumption by attachment to a fuel-burning appliance, to the fuel supply line thereto, or to the vent outlet or vent piping therefrom, shall not be installed unless *labeled* for such purpose and the installation is specifically *approved*.

SECTION 604
ELECTRICAL FACILITIES

604.1 Facilities required. Every occupied building shall be provided with an electrical system in compliance with the requirements of this section and Section 605.

604.2 Service. The size and usage of appliances and equipment shall serve as a basis for determining the need for additional facilities in accordance with NFPA 70. *Dwelling units* shall be served by a three-wire, 120/240 volt, single-phase electrical service having a minimum rating of 60 amperes.

604.3 Electrical system hazards. Where it is found that the electrical system in a *structure* constitutes a hazard to the *occupants* or the *structure* by reason of inadequate service, improper fusing, insufficient receptacle and lighting outlets, improper wiring or installation, *deterioration* or damage, or for similar reasons, the *code official* shall require the defects to be corrected to eliminate the hazard.

604.3.1 Abatement of electrical hazards associated with water exposure. The provisions of this section shall govern the repair and replacement of electrical systems and equipment that have been exposed to water.

604.3.1.1 Electrical equipment. Electrical distribution equipment, motor circuits, power equipment, transformers, wire, cable, flexible cords, wiring devices, ground fault circuit interrupters, surge protectors, molded case circuit breakers, low-voltage fuses, luminaires, ballasts, motors and electronic control, signaling and communication equipment that have been exposed to water shall be replaced in accordance with the provisions of the *International Building Code.*

Exception: The following equipment shall be allowed to be repaired where an inspection report from the equipment manufacturer or *approved* manufacturer's representative indicates that the equipment has not sustained damage that requires replacement:

1. Enclosed switches, rated not more than 600 volts or less.
2. Busway, rated not more than 600 volts.
3. Panelboards, rated not more than 600 volts.
4. Switchboards, rated not more than 600 volts.
5. Fire pump controllers, rated not more than 600 volts.
6. Manual and magnetic motor controllers.
7. Motor control centers.
8. Alternating current high-voltage circuit breakers.
9. Low-voltage power circuit breakers.
10. Protective relays, meters and current transformers.
11. Low- and medium-voltage switchgear.
12. Liquid-filled transformers.
13. Cast-resin transformers.
14. Wire or cable that is suitable for wet locations and whose ends have not been exposed to water.
15. Wire or cable, not containing fillers, that is suitable for wet locations and whose ends have not been exposed to water.
16. Luminaires that are listed as submersible.
17. Motors.
18. Electronic control, signaling and communication equipment.

604.3.2 Abatement of electrical hazards associated with fire exposure. The provisions of this section shall govern the repair and replacement of electrical systems and equipment that have been exposed to fire.

604.3.2.1 Electrical equipment. Electrical switches, receptacles and fixtures, including furnace, water heating, security system and power distribution circuits, that have been exposed to fire, shall be replaced in accordance with the provisions of the *International Building Code.*

Exception: Electrical switches, receptacles and fixtures that shall be allowed to be repaired where an inspection report from the equipment manufacturer or *approved* manufacturer's representative indicates that the equipment has not sustained damage that requires replacement.

SECTION 605
ELECTRICAL EQUIPMENT

605.1 Installation. Electrical equipment, wiring and appliances shall be properly installed and maintained in a safe and *approved* manner.

605.2 Receptacles. Every *habitable space* in a dwelling shall contain not less than two separate and remote receptacle outlets. Every laundry area shall contain not less than one grounding-type receptacle or a receptacle with a ground fault circuit interrupter. Every *bathroom* shall contain not less than one receptacle. Any new *bathroom* receptacle outlet shall have ground fault circuit interrupter protection. All receptacle outlets shall have the appropriate faceplate cover for the location.

605.3 Luminaires. Every public hall, interior stairway, *toilet room*, kitchen, *bathroom*, laundry room, boiler room and furnace room shall contain not less than one electric luminaire. Pool and spa luminaires over 15 V shall have ground fault circuit interrupter protection.

605.4 Wiring. Flexible cords shall not be used for permanent wiring, or for running through doors, windows, or cabinets, or concealed within walls, floors, or ceilings.

SECTION 606
ELEVATORS, ESCALATORS AND DUMBWAITERS

606.1 General. Elevators, dumbwaiters and escalators shall be maintained in compliance with ASME A17.1. The most current certificate of inspection shall be on display at all times within the elevator or attached to the escalator or dumbwaiter, be available for public inspection in the office of the building *operator* or be posted in a publicly conspicuous location *approved* by the *code official.* The inspection and tests shall be performed at not less than the periodic intervals listed in ASME A17.1, Appendix N, except where otherwise specified by the authority having jurisdiction.

606.2 Elevators. In buildings equipped with passenger elevators, not less than one elevator shall be maintained in operation at all times when the building is occupied.

> **Exception:** Buildings equipped with only one elevator shall be permitted to have the elevator temporarily out of service for testing or servicing.

SECTION 607
DUCT SYSTEMS

607.1 General. Duct systems shall be maintained free of obstructions and shall be capable of performing the required function.

CHAPTER 7
FIRE SAFETY REQUIREMENTS

User note:

About this chapter: *Chapter 7 establishes fire safety requirements for existing structures by providing requirements for means of egress, including path of travel, required egress width, means of egress doors and emergency escape openings, and for the maintenance of fire-resistance-rated assemblies, fire protection systems, and carbon monoxide alarm and detection systems.*

SECTION 701
GENERAL

701.1 Scope. The provisions of this chapter shall govern the minimum conditions and standards for fire safety relating to structures and exterior *premises*, including fire safety facilities and equipment to be provided.

701.2 Responsibility. The *owner* of the *premises* shall provide and maintain such fire safety facilities and equipment in compliance with these requirements. A person shall not occupy as *owner-occupant* or permit another person to occupy any *premises* that do not comply with the requirements of this chapter.

SECTION 702
MEANS OF EGRESS

[BE] 702.1 General. A safe, continuous and unobstructed path of travel shall be provided from any point in a building or *structure* to the *public way*. Means of egress shall comply with the *International Fire Code*.

[BE] 702.2 Aisles. The required width of aisles in accordance with the *International Fire Code* shall be unobstructed.

[BE] 702.3 Locked doors. Means of egress doors shall be readily openable from the side from which egress is to be made without the need for keys, special knowledge or effort, except where the door hardware conforms to that permitted by the *International Building Code*.

[BE] 702.4 Emergency escape and rescue openings. Required emergency escape and rescue openings shall be maintained in accordance with the code in effect at the time of construction, and both of the following:

1. Required emergency escape and rescue openings shall be operational from the inside of the room without the use of keys or tools.
2. Bars, grilles, grates or similar devices are permitted to be placed over emergency escape and rescue openings provided that the minimum net clear opening size complies with the code that was in effect at the time of construction and the unit is equipped with smoke alarms installed in accordance with Section 907.2.10 of the *International Building Code*. Such devices shall be releasable or removable from the inside without the use of a key, tool or force greater than that which is required for normal operation of the escape and rescue opening.

SECTION 703
FIRE-RESISTANCE RATINGS

[BF] 703.1 Fire-resistance-rated assemblies. The provisions of this chapter shall govern maintenance of the materials, systems and assemblies used for structural fire resistance and fire-resistance-rated construction separation of adjacent spaces to safeguard against the spread of fire and smoke within a building and the spread of fire to or from buildings.

[BF] 703.2 Unsafe conditions. Where any components are not maintained and do not function as intended or do not have the fire resistance required by the code under which the building was constructed or altered, such components or portions thereof shall be deemed unsafe conditions in accordance with Section 114.1.1 of the *International Fire Code*. Components or portions thereof determined to be unsafe shall be repaired or replaced to conform to that code under which the building was constructed or altered. Where the condition of components is such that any building, *structure* or portion thereof presents an *imminent danger* to the *occupants* of the building, *structure* or portion thereof, the fire *code official* shall act in accordance with Section 114.2 of the *International Fire Code*.

[BF] 703.3 Maintenance. The required fire-resistance rating of fire-resistance-rated construction, including walls, firestops, shaft enclosures, partitions, smoke barriers, floors, fire-resistive coatings and sprayed fire-resistant materials applied to structural members and joint systems, shall be maintained. Such elements shall be visually inspected annually by the *owner* and repaired, restored or replaced where damaged, altered, breached or penetrated. Records of inspections and repairs shall be maintained. Where concealed, such elements shall not be required to be visually inspected by the *owner* unless the concealed space is accessible by the removal or movement of a panel, access door, ceiling tile or entry to the space. Openings made therein for the passage of pipes, electrical conduit, wires, ducts, air transfer and any other reason shall be protected with *approved* methods capable of resisting the passage of smoke and fire. Openings through fire-resistance-rated assemblies shall be protected by self- or automatic-closing doors of *approved* construction meeting the fire protection requirements for the assembly.

[BF] 703.3.1 Fire blocking and draft stopping. Required fire blocking and draft stopping in combustible concealed spaces shall be maintained to provide continuity and integrity of the construction.

[BF] 703.3.2 Smoke barriers and smoke partitions. Required smoke barriers and smoke partitions shall be maintained to prevent the passage of smoke. Openings protected with *approved* smoke barrier doors or smoke dampers shall be maintained in accordance with NFPA 105.

[BF] 703.3.3 Fire walls, fire barriers, and fire partitions. Required fire walls, fire barriers and fire partitions shall be maintained to prevent the passage of fire. Openings protected with *approved* doors or fire dampers shall be maintained in accordance with NFPA 80.

[BF] 703.4 Opening protectives. Opening protectives shall be maintained in an operative condition in accordance with NFPA 80. The application of field-applied labels associated with the maintenance of opening protectives shall follow the requirements of the *approved* third-party certification organization accredited for listing the opening protective. Fire doors and smoke barrier doors shall not be blocked or obstructed, or otherwise made inoperable. Fusible links shall be replaced whenever fused or damaged. Fire door assemblies shall not be modified.

[BF] 703.4.1 Signs. Where required by the *code official,* a sign shall be permanently displayed on or near each fire door in letters not less than 1 inch (25 mm) high to read as follows:

1. For doors designed to be kept normally open: FIRE DOOR – DO NOT BLOCK.
2. For doors designed to be kept normally closed: FIRE DOOR – KEEP CLOSED.

[BF] 703.4.2 Hold-open devices and closers. Hold-open devices and automatic door closers shall be maintained. During the period that such a device is out of service for repairs, the door it operates shall remain in the closed position.

[BF] 703.4.3 Door operation. Swinging fire doors shall close from the full-open position and latch automatically. The door closer shall exert enough force to close and latch the door from any partially open position.

[BF] 703.5 Ceilings. The hanging and displaying of salable goods and other decorative materials from acoustical ceiling systems that are part of a fire-resistance-rated horizontal assembly shall be prohibited.

[BF] 703.6 Testing. Horizontal and vertical sliding and rolling fire doors shall be inspected and tested annually to confirm operation and full closure. Records of inspections and testing shall be maintained.

[BF] 703.7 Vertical shafts. Interior vertical shafts, including stairways, elevator hoistways and service and utility shafts, which connect two or more stories of a building shall be enclosed or protected as required in Chapter 11 of the *International Fire Code.* New floor openings in existing buildings shall comply with the *International Building Code.*

[BF] 703.8 Opening protective closers. Where openings are required to be protected, opening protectives shall be maintained self-closing or automatic-closing by smoke detection. Existing fusible-link-type automatic door-closing devices shall be replaced if the fusible link rating exceeds 135°F (57°C).

SECTION 704
FIRE PROTECTION SYSTEMS

[F] 704.1 Inspection, testing and maintenance. Fire protection and life safety systems shall be maintained in accordance with the *International Fire Code* in an operative condition at all times, and shall be replaced or repaired where defective.

[F] 704.1.1 Fire protection and life safety systems. Fire protection and life safety systems shall be installed, repaired, operated and maintained in accordance with this code the International Fire Code and the International Building Code.

[F] 704.1.2 Required fire protection and life safety systems. Fire protection and life safety systems required by this code, the *International Fire Code* or the *International Building Code* shall be installed, repaired, operated, tested and maintained in accordance with this code. A fire protection and life safety system for which a design option, exception or reduction to the provisions of this code, the *International Fire Code* or the *International Building Code* has been granted shall be considered to be a required system.

[F] 704.1.3 Fire protection systems. Fire protection systems shall be inspected, maintained and tested in accordance with the following *International Fire Code* requirements.

1. Automatic sprinkler systems, see Section 903.5.
2. Automatic fire-extinguishing systems protecting commercial cooking systems, see Section 904.13.5.
3. Automatic water mist extinguishing systems, see Section 904.11.
4. Carbon dioxide extinguishing systems, see Section 904.8.
5. Carbon monoxide alarms and carbon monoxide detection systems, see Section 915.6.
6. Clean-agent extinguishing systems, see Section 904.10.
7. Dry-chemical extinguishing systems, see Section 904.6.
8. Fire alarm and fire detection systems, see Section 907.8.
9. Fire department connections, see Sections 912.4 and 912.7.
10. Fire pumps, see Section 913.5.
11. Foam extinguishing systems, see Section 904.7.
12. Halon extinguishing systems, see Section 904.9.

13. Single- and multiple-station smoke alarms, see Section 907.10.
14. Smoke and heat vents and mechanical smoke removal systems, see Section 910.5.
15. Smoke control systems, see Section 909.22.
16. Wet-chemical extinguishing systems, see Section 904.5.

[F] 704.2 Standards. Fire protection systems shall be inspected, tested and maintained in accordance with the referenced standards listed in Table 704.2 and as required in this section.

TABLE 704.2
FIRE PROTECTION SYSTEM MAINTENANCE STANDARDS

SYSTEM	STANDARD
Portable fire extinguishers	NFPA 10
Carbon dioxide fire-extinguishing system	NFPA 12
Halon 1301 fire-extinguishing systems	NFPA 12A
Dry-chemical extinguishing systems	NFPA 17
Wet-chemical extinguishing systems	NFPA 17A
Water-based fire protection systems	NFPA 25
Fire alarm systems	NFPA 72
Smoke and heat vents	NFPA 204
Water-mist systems	NFPA 750
Clean-agent extinguishing systems	NFPA 2001

[F] 704.2.1 Records. Records shall be maintained of all system inspections, tests and maintenance required by the referenced standards.

[F] 704.2.2 Records information. Initial records shall include the: name of the installation contractor; type of components installed; manufacturer of the components; location and number of components installed per floor; and manufacturers' operation and maintenance instruction manuals. Such records shall be maintained for the life of the installation.

[F] 704.3 Systems out of service. Where a required fire protection system is out of service, the fire department and the fire *code official* shall be notified immediately and, where required by the fire *code official*, either the building shall be evacuated or an *approved* fire watch shall be provided for all *occupants* left unprotected by the shutdown until the fire protection system has been returned to service. Where utilized, fire watches shall be provided with not less than one *approved* means for notification of the fire department and shall not have duties beyond performing constant patrols of the protected *premises* and keeping watch for fires. Actions shall be taken in accordance with Section 901 of the *International Fire Code* to bring the systems back in service.

Exception: Facilities with an approved notification and impairment management program. The notification and impairment program for water-based fire protection systems shall comply with NFPA 25.

[F] 704.3.1 Emergency impairments. Where unplanned impairments of fire protection systems occur, appropriate emergency action shall be taken to minimize potential injury and damage. The impairment coordinator shall implement the steps outlined in Section 901.7.4 of the *International Fire Code.*

[F] 704.4 Removal of or tampering with equipment. It shall be unlawful for any person to remove, tamper with or otherwise disturb any fire protection or life safety system required by this code except for the purposes of extinguishing fire, training, recharging or making necessary repairs.

[F] 704.4.1 Removal of or tampering with appurtenances. Locks, gates, doors, barricades, chains, enclosures, signs, tags and seals that have been installed by or at the direction of the fire *code official* shall not be removed, unlocked, destroyed or tampered with in any manner.

[F] 704.4.2 Removal of existing occupant-use hose lines. The fire *code official* is authorized to permit the removal of existing *occupant*-use hose lines where all of the following apply:

1. The installation is not required by the *International Fire Code* or the *International Building Code.*
2. The hose line would not be utilized by trained personnel or the fire department.
3. The remaining outlets are compatible with local fire department fittings.

[F] 704.4.3 Termination of monitoring service. For fire alarm systems required to be monitored by the *International Fire Code*, notice shall be made to the fire *code official* whenever alarm monitoring services are terminated. Notice shall be made in writing by the provider of the monitoring service being terminated.

[F] 704.5 Fire department connection. Where the fire department connection is not visible to approaching fire apparatus, the fire department connection shall be indicated by an *approved* sign mounted on the street front or on the side of the building. Such sign shall have the letters "FDC" not less than 6 inches (152 mm) high and words in letters not less than 2 inches (51 mm) high or an arrow to indicate the location. Such signs shall be subject to the approval of the fire *code official.*

[F] 704.5.1 Fire department connection access. Ready access to fire department connections shall be maintained at all times and without obstruction by fences, bushes, trees, walls or any other fixed or movable object. Access to fire department connections shall be *approved* by the fire chief.

Exception: Fences, where provided with an access gate equipped with a sign complying with the legend requirements of Section 912.5 of the *International Fire Code* and a means of emergency operation. The gate and the means of emergency operation shall be *approved* by the fire chief and maintained operational at all times.

[F] 704.5.2 Clear space around connections. A working space of not less than 36 inches (914 mm) in width, 36 inches (914 mm) in depth and 78 inches (1981 mm) in height shall be provided and maintained in front of and to the sides of wall-mounted fire department connections and around the circumference of free-standing fire department connections.

[F] 704.6 Single- and multiple-station smoke alarms. Single- and multiple-station smoke alarms shall be installed in existing Group I-1 and R *occupancies* in accordance with Sections 704.6.1 through 704.6.3.

[F] 704.6.1 Where required. Existing Group I-1 and R *occupancies* shall be provided with single-station smoke alarms in accordance with Sections 704.6.1.1 through 704.6.1.4. Interconnection and power sources shall be in accordance with Sections 704.6.2 and 704.6.3.

Exceptions:

1. Where the code that was in effect at the time of construction required smoke alarms and smoke alarms complying with those requirements are already provided.
2. Where smoke alarms have been installed in occupancies and dwellings that were not required to have them at the time of construction, additional smoke alarms shall not be required provided that the existing smoke alarms comply with requirements that were in effect at the time of installation.
3. Where smoke detectors connected to a fire alarm system have been installed as a substitute for smoke alarms.

[F] 704.6.1.1 Group R-1. Single- or multiple-station smoke alarms shall be installed in all of the following locations in Group R-1:

1. In sleeping areas.
2. In every room in the path of the means of egress from the sleeping area to the door leading from the *sleeping unit*.
3. In each story within the *sleeping unit*, including *basements*. For *sleeping units* with split levels and without an intervening door between the adjacent levels, a smoke alarm installed on the upper level shall suffice for the adjacent lower level provided that the lower level is less than one full story below the upper level.

[F] 704.6.1.2 Groups R-2, R-3, R-4 and I-1. Single or multiple-station smoke alarms shall be installed and maintained in Groups R-2, R-3, R-4 and I-1 regardless of *occupant load* at all of the following locations:

1. On the ceiling or wall outside of each separate sleeping area in the immediate vicinity of bedrooms.
2. In each room used for sleeping purposes.
3. In each story within a *dwelling unit*, including *basements* but not including crawl spaces and uninhabitable attics. In *dwellings* or *dwelling units* with split levels and without an intervening door between the adjacent levels, a smoke alarm installed on the upper level shall suffice for the adjacent lower level provided that the lower level is less than one full story below the upper level.

[F] 704.6.1.3 Installation near cooking appliances. Smoke alarms shall not be installed in the following locations unless this would prevent placement of a smoke alarm in a location required by Section 704.6.1.1 or 704.6.1.2.

1. Ionization smoke alarms shall not be installed less than 20 feet (6096 m) horizontally from a permanently installed cooking appliance.
2. Ionization smoke alarms with an alarm-silencing switch shall not be installed less than 10 feet (3048 mm) horizontally from a permanently installed cooking appliance.
3. Photoelectric smoke alarms shall not be installed less than 6 feet (1829 mm) horizontally from a permanently installed cooking appliance.

[F] 704.6.1.4 Installation near bathrooms. Smoke alarms shall be installed not less than 3 feet (914 mm) horizontally from the door or opening of a *bathroom* that contains a bathtub or shower unless this would prevent placement of a smoke alarm required by Section 704.6.1.1 or 704.6.1.2.

[F] 704.6.2 Interconnection. Where more than one smoke alarm is required to be installed within an individual *dwelling* or *sleeping unit*, the smoke alarms shall be interconnected in such a manner that the activation of one alarm will activate all of the alarms in the individual unit. Physical interconnection of smoke alarms shall not be required where listed wireless alarms are installed and all alarms sound upon activation of one alarm. The alarm shall be clearly audible in all bedrooms over background noise levels with all intervening doors closed.

Exceptions:

1. Interconnection is not required in buildings that are not undergoing *alterations*, repairs or construction of any kind.
2. Smoke alarms in existing areas are not required to be interconnected where *alterations* or repairs do not result in the removal of interior wall or ceiling finishes exposing the structure, unless there is an attic, crawl space or *basement* available that could provide access for interconnection without the removal of interior finishes.

[F] 704.6.3 Power source. Single-station smoke alarms shall receive their primary power from the building wiring provided that such wiring is served from a commercial source and shall be equipped with a battery backup. Smoke alarms with integral strobes that are not equipped with battery backup shall be connected to an emergency electrical system. Smoke alarms shall emit a signal when the batteries are low. Wiring shall be permanent and without a disconnecting switch other than as required for overcurrent protection.

Exceptions:

1. Smoke alarms are permitted to be solely battery operated in existing buildings where construction is not taking place.
2. Smoke alarms are permitted to be solely battery operated in buildings that are not served from a commercial power source.
3. Smoke alarms are permitted to be solely battery operated in existing areas of buildings undergoing *alterations* or repairs that do not result in the removal of interior walls or ceiling finishes exposing the structure, unless there is an attic, crawl space or *basement* available that could provide access for building wiring without the removal of interior finishes.

[F] 704.6.4 Smoke detection system. Smoke detectors listed in accordance with UL 268 and provided as part of the building's fire alarm system shall be an acceptable alternative to single- and multiple-station smoke alarms and shall comply with the following:

1. The fire alarm system shall comply with all applicable requirements in Section 907 of the *International Fire Code.*
2. Activation of a smoke detector in a dwelling or sleeping unit shall initiate alarm notification in the *dwelling* or *sleeping unit* in accordance with Section 907.5.2 of the *International Fire Code.*
3. Activation of a smoke detector in a *dwelling* or *sleeping unit* shall not activate alarm notification appliances outside of the *dwelling* or *sleeping unit*, provided that a supervisory signal is generated and monitored in accordance with Section 907.6.6 of the *International Fire Code.*

[F] 704.7 Single- and multiple-station smoke alarms. Single- and multiple-station smoke alarms shall be tested and maintained in accordance with the manufacturer's instructions. Smoke alarms that do not function shall be replaced. Smoke alarms installed in one- and two-family dwellings shall be replaced not more than 10 years from the date of manufacture marked on the unit, or shall be replaced if the date of manufacture cannot be determined.

SECTION 705
CARBON MONOXIDE ALARMS AND DETECTION

[F] 705.1 General. Carbon monoxide alarms shall be installed in dwellings in accordance with Section 1103.9 of the *International Fire Code*, except that alarms in dwellings covered by the *International Residential Code* shall be installed in accordance with Section R315 of that code.

[F] 705.2 Carbon monoxide alarms and detectors. Carbon monoxide alarms and carbon monoxide detection systems shall be maintained in accordance with NFPA 720. Carbon monoxide alarms and carbon monoxide detectors that become inoperable or begin producing end-of-life signals shall be replaced.

CHAPTER 8
REFERENCED STANDARDS

User note:

About this chapter: *This code contains numerous references to standards promulgated by other organizations that are used to provide requirements for materials and methods of construction. Chapter 8 contains a comprehensive list of all standards that are referenced in this code. These standards, in essence, are part of this code to the extent of the reference to the standard.*

This chapter lists the standards that are referenced in various sections of this document. The standards are listed herein by the promulgating agency of the standard, the standard identification, the effective date and title and the section or sections of this document that reference the standard. The application of the referenced standards shall be as specified in Section 102.8.

ASME

American Society of Mechanical Engineers
Two Park Avenue
New York, NY 10016-5990

ASME A17.1—2019/CSA B44—19: **Safety Code for Elevators and Escalators**
606.1

ASTM

ASTM International
100 Barr Harbor Drive, P.O. Box C700
West Conshohocken, PA 19428-2959

F1346—91 (2018): **Performance Specifications for Safety Covers and Labeling Requirements for All Covers for Swimming Pools, Spas and Hot Tubs**
303.2

ICC

International Code Council
500 New Jersey Avenue, NW 6th Floor
Washington, DC 20001

IBC—21: **International Building Code®**
102.3, 201.3, 304.1.1, 305.1.1, 306.1.1, 403.1, 604.3.1.1, 604.3.2.1, 702.3, 704.4.2

IEBC—21: **International Existing Building Code®**
102.3, 201.3, 304.1.1, 305.1.1, 306.1.1

IECC—21: **International Energy Conservation Code®**
102.3

IFC—21: **International Fire Code®**
102.3, 201.3, 604.3.1.1, 702.1, 702.2, 704.1, 704.1.2, 704.1.3, 704.3, 704.3.1, 704.4.2, 704.4.3, 704.5.1, 704.6.4, 705.1

IFGC—21: **International Fuel Gas Code®**
102.3, 201.3

IMC—21: **International Mechanical Code®**
102.3, 201.3

IPC—21: **International Plumbing Code®**
102.3, 201.3, 502.5, 505.1, 505.5.1, 602.2, 602.3

IRC—21: **International Residential Code®**
102.3, 201.3

IZC—21: **International Zoning Code®**
102.3, 201.3

NFPA

National Fire Protection Association
1 Batterymarch Park
Quincy, MA 02169-7471

10—21: Standard for Portable Fire Extinguishers
Table 704.2

12—18: Standard on Carbon Dioxide Extinguishing Systems
Table 704.2

12A—18: Standard on Halon 1301 Fire Extinguishing Systems
Table 704.2

17—20: Standard for Dry Chemical Extinguishing Systems
Table 704.2

17A—20: Standard for Wet Chemical Extinguishing Systems
Table 704.2

25—20: Standard for the Inspection, Testing and Maintenance of Water-Based Fire Protection Systems
Table 704.2

70—20: National Electrical Code
102.3, 201.3, 604.2

72—19: National Fire Alarm and Signaling Code
Table 704.2

80—19: Standard for Fire Doors and Other Opening Protectives
703.3.3, 703.4

105—19: Standard for Smoke Door Assemblies and Other Opening Protectives
703.3.2

204—18: Standard for Smoke and Heat Venting
Table 704.2

720—15: Standard for the Installation of Carbon Monoxide (CO) Detection and Warning Equipment
705.2

750—19: Standard on Water Mist Fire Protection Systems
Table 704.2

2001—18: Standard on Clean Agent Fire Extinguishing Systems
Table 704.2

UL

Underwriters Laboratories, LLC
333 Pfingsten Road
Northbrook, IL 60062

268—2016: Smoke Detectors for Fire Alarm Systems—with revisions through July 2016
704.6.4

APPENDIX A
BOARDING STANDARD

The provisions contained in this appendix are not mandatory unless specifically referenced in the adopting ordinance.

User note:

About this appendix: *Appendix A provides minimum specifications for boarding a structure. This can be utilized by a jurisdiction as a set of minimum requirements in order to result in consistent boarding quality. These requirements also provide a reasonable means to eliminate having to approve numerous methods or materials for the boarding and securing of a structure. It is important to note that the provisions of Appendix A are not mandatory unless specifically referenced in the adopting ordinance of the authority having jurisdiction.*

A101 GENERAL

A101.1 General. Windows and doors shall be boarded in an *approved* manner to prevent entry by unauthorized persons and shall be painted to correspond to the color of the existing *structure.*

A102 MATERIALS

A102.1 Boarding sheet material. Boarding sheet material shall be minimum $^1/_2$-inch-thick (12.7 mm) wood structural panels complying with the *International Building Code.*

A102.2 Boarding framing material. Boarding framing material shall be minimum nominal 2-inch by 4-inch (51 mm by 102 mm) solid sawn lumber complying with the *International Building Code.*

A102.3 Boarding fasteners. Boarding fasteners shall be minimum $^3/_8$-inch-diameter (9.5 mm) carriage bolts of such a length as required to penetrate the assembly and as required to adequately attach the washers and nuts. Washers and nuts shall comply with the *International Building Code.*

A103 INSTALLATION

A103.1 Boarding installation. The boarding installation shall be in accordance with Figures A103.1(1) and A103.1(2) and Sections A103.2 through A103.5.

A103.2 Boarding sheet material. The boarding sheet material shall be cut to fit the door or window opening neatly or shall be cut to provide an equal overlap at the perimeter of the door or window.

A103.3 Windows. The window shall be opened to allow the carriage bolt to pass through or the window sash shall be removed and stored. The 2-inch by 4-inch (51 mm by 102 mm) strong back framing material shall be cut minimum 2 inches (51 mm) wider than the window opening and shall be placed on the inside of the window opening 6 inches (152 mm) minimum above the bottom and below the top of the window opening. The framing and boarding shall be predrilled. The assembly shall be aligned and the bolts, washers and nuts shall be installed and secured.

A103.4 Door walls. The door opening shall be framed with minimum 2-inch by 4-inch (51 mm by 102 mm) framing material secured at the entire perimeter and vertical members at a maximum of 24 inches (610 mm) on center. Blocking shall also be secured at a maximum of 48 inches (1219 mm) on center vertically. Boarding sheet material shall be secured with screws and nails alternating every 6 inches (152 mm) on center.

A103.5 Doors. Doors shall be secured by the same method as for windows or door openings. One door to the *structure* shall be available for authorized entry and shall be secured and locked in an *approved* manner.

A104 REFERENCED STANDARD

A104.1 Referenced standards. Table A104.1 lists the standard that is referenced in various sections of this appendix. The standard is listed herein by the standard identification, the effective date and title and the section or sections of this document that reference the standard. The application of the referenced standards shall be as specified in Section 102.8.

TABLE A104.1
REFERENCED STANDARD

STANDARD ACRONYM	STANDARD NAME	SECTIONS HEREIN REFERENCED
IBC—21	International Building Code	A102.1, A102.2, A102.3

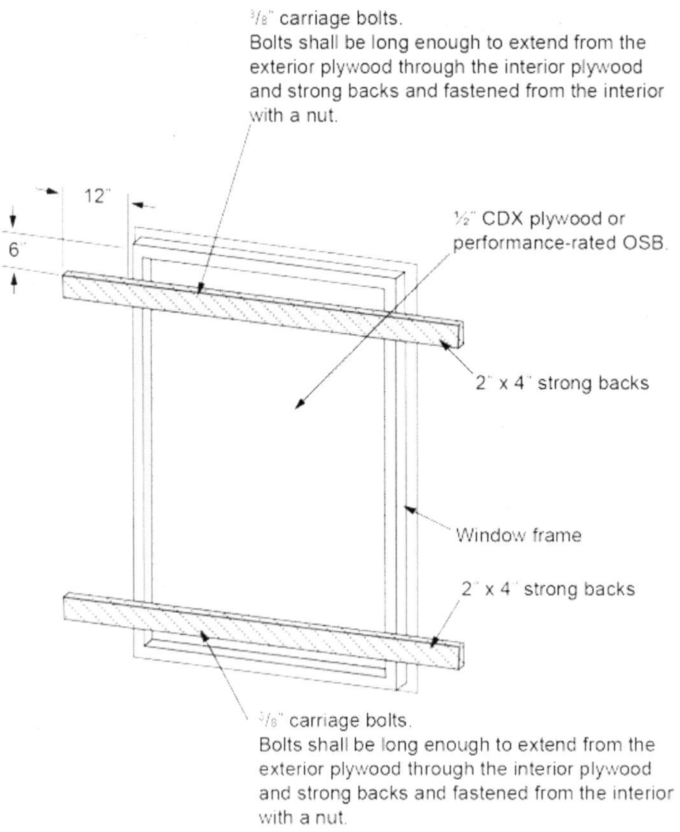

For SI: 1 inch = 25.4 mm.

FIGURE A103.1(1)
BOARDING OF DOOR OR WINDOW

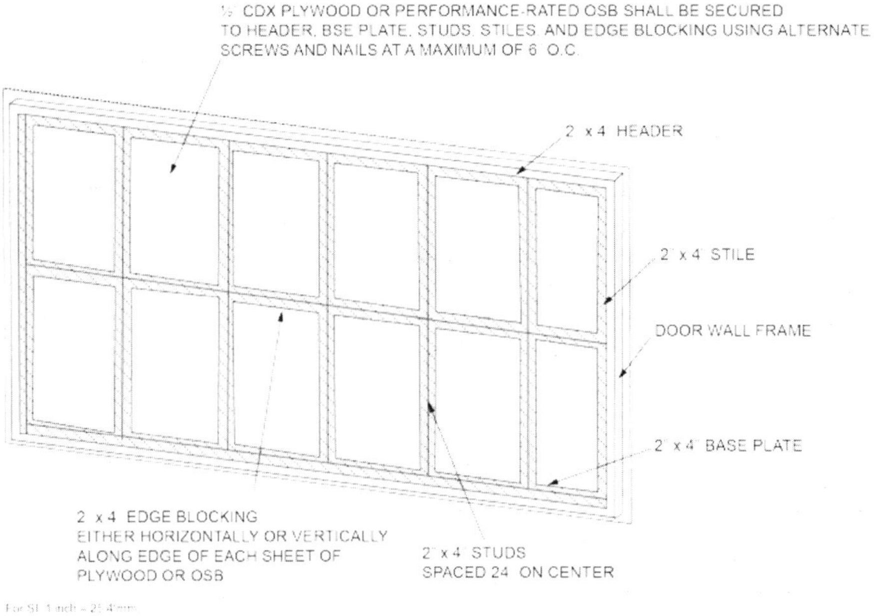

FIGURE A103.1(2)
BOARDING OF DOOR WALL

APPENDIX B
BOARD OF APPEALS

The provisions contained in this appendix are not mandatory unless specifically referenced in the adopting ordinance.

User note:

About this appendix: *Appendix B provides criteria for Board of Appeals members. Also provided are procedures by which the Board of Appeals should conduct its business.*

Code development reminder: *Code change proposals to this appendix will be considered by the Administrative Code Development Committee during the 2022 (Group B) Code Development Cycle.*

SECTION B101
GENERAL

B101.1 Scope. A board of appeals shall be established within the jurisdiction for the purpose of hearing applications for modification of the requirements of this code pursuant to the provisions of Section 107 (Means of Appeals). The board shall be established and operated in accordance with this section, and shall be authorized to hear evidence from appellants and the code official pertaining to the application and intent of this code for the purpose of issuing orders pursuant to these provisions.

B101.2 Application for appeal. Any person shall have the right to appeal a decision of the *code official* to the board. An application for appeal shall be based on a claim that the intent of this code or the rules legally adopted hereunder have been incorrectly interpreted, the provisions of this code do not fully apply or an equally good or better form of construction is proposed. The application shall be filed on a form obtained from the *code official* within 20 days after the notice was served.

B101.2.1 Limitation of authority. The board shall not have authority to waive requirements of this code or interpret the administration of this code.

B101.2.2 Stays of enforcement. Appeals of notice and orders, other than *Imminent Danger* notices, shall stay the enforcement of the notice and order until the appeal is heard by the board.

**

B101.3 Membership of the board. The board shall consist of five voting members appointed by the chief appointing authority of the jurisdiction. Each member shall serve for [INSERT NUMBER OF YEARS] years or until a successor has been appointed. The board member's terms shall be staggered at intervals, so as to provide continuity. The *code official* shall be an ex officio member of said board but shall not vote on any matter before the board.

B101.3.1 Qualifications. The board shall consist of five individuals, who are qualified by experience and training to pass on matters pertaining to building construction and are not employees of the jurisdiction.

B101.3.2 Alternate members. The chief appointing authority is authorized to appoint two alternate members who shall be called by the board chairperson to hear appeals during the absence or disqualification of a member. Alternate members shall possess the qualifications required for board membership, and shall be appointed for the same term or until a successor has been appointed.

B101.3.3 Vacancies. Vacancies shall be filled for an unexpired term in the same manner in which original appointments are required to be made.

B101.3.4 Chairperson. The board shall annually select one of its members to serve as chairperson.

B101.3.5 Secretary. The chief appointing authority shall designate a qualified clerk to serve as secretary to the board. The secretary shall file a detailed record of all proceedings which shall set forth the reasons for the board's decision, the vote of each member, the absence of a member and any failure of a member to vote.

B101.3.6 Conflict of interest. A member with any personal, professional or financial interest in a matter before the board shall declare such interest and refrain from participating in discussions, deliberations and voting on such matters.

B101.3.7 Compensation of members. Compensation of members shall be determined by law.

B101.3.8 Removal from the board. A member shall be removed from the board prior to the end of their terms only for cause. Any member with continued absence from regular meeting of the board may be removed at the discretion of the chief appointing authority.

B101.4 Rules and procedures. The board shall establish policies and procedures necessary to carry out its duties consistent with the provisions of this code and applicable state law. The procedures shall not require compliance with strict rules of evidence, but shall mandate that only relevant information be presented.

B101.5 Notice of meeting. The board shall meet upon notice from the chairperson, within 10 days of the filing of an appeal or at stated periodic intervals.

B101.5.1 Open hearing. All hearings before the board shall be open to the public. The appellant, the appellant's representative, the *code official* and any person whose interests are affected shall be given an opportunity to be heard.

B101.5.2 Quorum. Three members of the board shall constitute a quorum.

B101.5.3 Postponed hearing. When five members are not present to hear an appeal, either the appellant or the appellant's representative shall have the right to request a postponement of the hearing.

B101.6 Legal counsel. The jurisdiction shall furnish legal counsel to the board to provide members with general legal advice concerning matters before them for consideration. Members shall be represented by legal counsel at the jurisdiction's expense in all matters arising from service within the scope of their duties.

B101.7 Board decision. The board shall only modify or reverse the decision of the *code official* by a concurring vote of three or more members.

B101.7.1 Resolution. The decision of the board shall be by resolution. Every decision shall be promptly filed in writing in the office of the *code official* within three days and shall be open to the public for inspection. A certified copy shall be furnished to the appellant or the appellant's representative and to the *code official.*

B101.7.2 Administration. The *code official* shall take immediate action in accordance with the decision of the board.

B101.8 Court review. Any person, whether or not a previous party of the appeal, shall have the right to apply to the appropriate court for a writ of certiorari to correct errors of law. Application for review shall be made in the manner and time required by law following the filing of the decision in the office of the chief administrative officer.

INDEX

C

D

E

W

Y